ENERGY AND HUMAN WELFARE—
A CRITICAL ANALYSIS

Volume II
**Alternative Technologies
for Power Production**

ENERGY AND HUMAN WELFARE—A CRITICAL ANALYSIS

A Selection of Papers on the Social, Technological, and Environmental Problems of Electric Power Consumption

Edited by
BARRY COMMONER
HOWARD BOKSENBAUM
MICHAEL CORR

Prepared for the Electric Power Task Force of the Scientists' Institute for Public Information and the Power Study Group of the American Association for the Advancement of Science Committee on Environmental Alterations

VOLUME II

Alternative Technologies for Power Production

MACMILLAN INFORMATION
A Division of Macmillan Publishing Co., Inc.
New York

Macmillan Information
A Division of Macmillan Publishing Co., Inc.
866 Third Avenue, New York, N.Y. 10022

Collier-Macmillan Canada Ltd.

Library of Congress Catalog Card Number: 75–8987

Printed in the United States of America

printing number

1 2 3 4 5 6 7 8 9 10

Library of Congress Cataloging in Publication Data
Main entry under title:

Alternative technologies for power production.

    (Energy and human welfare ; v. 2)
    Includes bibliographies and index.
    1. Fuel--Addresses, essays, lectures. 2. Electric power production--Addresses, essays, lectures.
3. Power resources--Addresses, essays, lectures.
I. Commoner, Barry, 1917-   II. Boksenbaum,
Howard. III. Corr, Michael. IV. Scientists'
Institute for Public Information. Electric Power
Task Force. V. American Association for the Advancement of Science. Committee on Environmental
Alterations. Power Study Group. VI. Series.
TP319.A56   662'.6   75-8987
ISBN 0-02-468430-9

Contents

STATEMENT OF THE BOARD
OF DIRECTORS OF THE AMERICAN
ASSOCIATION FOR THE
ADVANCEMENT OF SCIENCE

The following expression of the Board's Statement of Policy concerning AAAS Committee Activities and Reports: "Responsibility for statements of fact and expressions of opinion contained in this report rests with the committee that prepared it.* The AAAS Board of Directors, in accordance with Association policy and without passing judgment on the views expressed, has approved its publication as a contribution to the discussion of an important issue."

*Relevant to this statement of policy, the Committee on Environmental Alterations commissioned the Power Study Working Group to select the authors for the preparation of this report. The contributors to the report are responsible for the statements of fact and expressions of opinion in their respective papers.

Members of the AAAS-CEA who participated in organizing these volumes and their affiliation at that time:

Barry Commoner, Washington University
Donald Aitken, San Jose State College
F. Herbert Borman, Yale University
Theodore Byerly, U.S. Department of Agriculture
William Capron, Harvard University
H. Jack Geiger, SUNY at Stony Brook
Oscar Harkavy, Ford Foundation
Walter Modell, College of Medicine, Cornell University
Arthur Squires, City University of New York

STATEMENT OF THE COMMITTEE
ON ENVIRONMENTAL ALTERATIONS

It has become clear that the majority of the people of the United States are working to stem the escalation of environmental problems resulting from the incessant expansion of industrial technology. It is this Committee's purpose to provide the public with an assessment of the environmental consequences of our technology.

Energy-related environmental problems and their associated social issues are particularly pressing since the energy industry and, in particular, the electric power industry, appear to be a major force in our economy. Lead times for major decisions in the electric power industry are of the order of 10 to 25 years, which makes it difficult for the public, the news media, and policy makers to engage in timely, well-informed debate over the crucial environmental and social problems created by increasing energy consumption.

To help fill the need for documentation on these problems, the Committee on Environmental Alterations, in collaboration with the Scientists' Institute for Public Information, established a Task Force to prepare a report on the environmental impact of power production and possible means of alleviating adverse effects. One result of the work of this Task Force has been the preparation of these three volumes. In the view of the Committee, this is a useful contribution to our understanding of this complex and urgent problem.

We recognize that each of the authors of the separate papers that make up the main part of the report have strong, well-developed views on their subjects which may not be shared by their colleagues. This is to be expected in a field which is still poorly understood. Where sharp differences of approach to a given problem exist, the report includes, if only in brief form, some material indicative of the divergence in views.

The authors of each article take full responsibility for their contributions and, on its part, the Committee on Environmental Alterations hopes that members of the scientific community and of the public at large will find this compendium a useful contribution to an increasingly urgent subject.

The Committee wishes to express its particular appreciation to Professor Dean Abrahamson and to the Task Force members for their efforts in providing this valuable document.

Acknowledgements

The editors wish to express their gratitude to all of the authors whose cooperation was so important to the production of this volume: Special thanks go to Artis Bernard for her help in resolving questions of style. Equally important was the efficient and expert help of those who contributed to the preparation of the manuscript: the artwork was done by Dru Lipsitz and Lurline Hogsett; typing and maintenance of mail communications were handled by Gladys Yandell, Peggy Whitton, Amy Papian, Lenore Harris, Jane Murdock, Sandy Marshall, Leslie Rutlin, Pieriette Murray, Cynthia Glastris and Julie Love.

We would also like to thank J. Klarmann, K. H. Lusczynski, Michael Friedlander and Caleb A. Smith for their assistance.

In addition, we wish to acknowledge the cooperation of various publishers in granting us permission to reproduce those papers that have also appeared elsewhere. Certain chapters have appeared as articles in *Environment* magazine, and are copyrighted in the indicated year by the Scientists' Institute for Public Information:

"Black Box," December 1971

Chapter 3 in a revised form appeared in *American Institute of Architecture Journal*, June 1972.

Chapter 6 appeared as "Controlled Nuclear Fusion: Status and Outlook," in *Science*, May 21, 1971, Volume 172, pp. 797-808.

Preface

The urgent need for a national energy policy is now widely recognized. Such a policy can succeed only if it takes into account the basic facts about the different ways of producing and using energy. This volume considers the problems of producing energy; Volume III takes up how it is used.

Modern society has created an elaborate, rapidly growing energy system, motivated by the desire to gain the economic value that energy can yield. Energy itself has no use value; such value is created only when energy is converted into an economic or social good: a warm home, the movement of freight or passengers, the conversion of ore into metal; the fashioning of metal into useful objects. Moreover, energy is not really "produced"; what we actually do is to transform energy that is available in some form in nature into more useful forms.

There are, therefore, two basic requirements for an energy system: (a) a natural source of energy such as fuel and (b) the machinery needed to transform the original form of energy into a more useful form—for example, a furnace that converts energy latent in the molecules of coal or oil into heat, or a power plant that converts some of the nuclear matter of uranium atoms into energy. In general, the characteristics of the machinery reflect the properties of the natural energy source.

There are a variety of natural energy sources: oil and gas, coal, uranium and other nuclear fuels, geothermal energy and solar energy (which includes hydroelectric power, since the water that falls from a dam is lifted, to begin with, by the sun). Apart from their relative environmental effects (which are considered in Vol. I), these energy sources differ considerably in several important respects, which strongly influence their relative ability to produce the sought-for benefits.

1. Renewability.   Oil, gas and coal are fossil fuels, laid down only once in the earth's geological history as the residue of the photosynthetic activity of long-extinct plants. The amounts of these fuels beneath the earth's surface are therefore fixed and the amounts remaining for future use become smaller as the fuels are consumed. Naturally not all the fossil fuel that exists on the earth can be reached, or even found. Even the most optimistic estimate of the known deposits represents only a very small fraction of the amount of photosynthetic product that must have accumulated in the period when the fossil fuels were laid down.

Based on what is known about potentially accessible deposits of the fossil fuels, we can make very rough estimates of how long their supplies will last at the expected rates of consumption. In general, gas and crude oil can be expected to last some decades—perhaps 30–50 years or more—while coal would last for some centuries—300–500 years or more. (See Chapter 1) Thus, viewed on a scale of human history that is (optimistically) long-term, fossil fuels represent only a temporary source of energy.

When nuclear energy was first discovered, many people, impressed with the enormous energy content of a small amount of such fuel (one ounce of uranium 235 can yield as much energy as 85 tons of coal), regarded nuclear energy as the answer to our long-term needs. As shown in Chapter 2, this view is mistakenly optimistic; uranium ore of the richness that is now used commercially will probably be depleted in 25–50 years, at the present rate of use. Indeed, it is now apparent that unless the Liquid Metal Fast Breeder Reactor (LMFBR) is brought into operation, the present generation of nuclear power plants will run out of fuel at the same rate as plants that use oil or gas. The LMFBR is being developed because, by generating plutonium as it operates, it can extend present nuclear fuel resources considerably, perhaps as long as coal supplies.

There is a sharp distinction between all these sources of energy and solar energy: Solar energy is renewable, whereas, as we have seen, all other fuels, whether fossil or nuclear, are limited in amount and are simply exhausted as energy is released from them and used. (Another difference is that only solar energy is essentially free of environmental impact. As made evident in Vol. I, all other fuels despoil the land as they are extracted and pollute the environment with either chemical or radioactive contaminants as they are used. In contrast, the acquisition of solar energy involves only the transfer of energy, absorbed from sunlight, from one place on the earth's surface to another. This is a process that occurs naturally with considerable daily and seasonal variation, in the form of weather. The capture of enough solar energy to supply all U.S. needs would cause perturbations that would probably be small enough, relative to the natural fluctuations of the weather, to have no noticeable environmental effect.)

2. Conversion Technology.   The various natural energy sources also differ with respect to the kind of machinery needed to release the energy. Among the fossil fuels, gas requires relatively simple technology: a drilling rig taps an underground reservoir of gas, which is then driven by pumps through pipelines to where it is needed; there it is stored, further distributed by a system of pipes and finally used by the simple expedient of setting it ablaze and collecting the resultant heat. For rural use, a gaseous fuel is liquified by compression, shipped in that form and used, as gas, by the simple expedient of releasing it gradually through a pressure-reducing valve. Little can be done to advance this kind of technology, either to improve the efficiency of conversion to heat, or to reduce its already low impact on the environment.

In the case of coal, the technology of mining is relatively simple and straightforward and with two very important exceptions, probably not subject to much improvement. Two urgently needed improvements are better means of ensuring the miners' safety and health and better methods to reduce environmental impact, especially of strip-mining. In contrast, the technology used at present to release energy from coal is still primitive, and a number of improvements are in sight. Chapters 3 and 4 describe the increasing effort to find more efficient ways of burning coal with reduced environmental impact. Other research on coal technology involves the development of methods of converting coal to gaseous and liquid fuels by chemical hydrogenation. The latter process may become increasingly important as oil supplies become scarce, for liquid hydrocarbon fuel is essential for the operation of the vehicles that dominate our present transportation system—automobiles, trucks, and airplanes. However, this process (and the similar effort to produce oil from deposits of oil shale) involves serious environmental hazards, which are yet to be mastered.

In contrast with fossil fuels, the techniques involved in producing nuclear energy are far more complex, and yet in urgent need of improvements which will ultimately render them even more complex. The mining of uranium is a serious hazard to the miners, who run the risk of an elevated incidence of cancer from the attendant radiation. Efforts to reduce this hazard would require the addition of considerable technological complexities, such as extensive ventilation and protective equipment. Fossil fuels can be used to produce electricity or other forms of energy with minimal pretreatment such as distillation of crude oil to produce a fuel oil fraction, or even none (as in the case of gas or coal). In sharp contrast, nuclear fuel must be elaborately purified from uranium ore. The processing plants that prepare nuclear fuels are among the largest and most complex industrial operations ever developed. They consume considerable quantities of fossil fuel and involve significant environmental hazards, for example, from leakage of radioactive wastes. Nuclear reactors are, of course, themselves based on very complex and—compared with fossil fuels—as yet unreliable technology.

Because of the intense radiation involved in nuclear power production, nuclear reactors require numerous controls and complex safety devices. The unreliability of nuclear reactors is in part due to this complexity and in part to our still-limited experience with the effects of radiation. Thus in 1974, unexpected cracks in fuel rods (apparently an effect of radiation) forced the shutdown of a number of reactors. The LMFBR is even more complex and unproven than current conventional reactors. Its technological complexity is evident in the doubling of the anticipated cost of the first demonstration reactor within a year after the initial estimates were made. Again in contrast with fossil fuels, the residue of spent nuclear fuels cannot be dealt with without the intervention of very complicated and exceedingly unreliable technology.

An example of just how unreliable this technology is at present is evident from the dismal experience with the new fuel recovery plant at Morris, Illinois. Built at a cost of $64 million, the plant, which was supposed to open in 1971, has now been found to be useless, apparently because "the technology used in the plant does not work," (St. Louis Globe-Democrat, Aug. 29, 1974). The plant may have to be scrapped and a new one built at a cost estimated at between $90 and $130 million. Finally, we should take note that no satisfactory means has as yet been worked out to safely dispose of the highly radioactive wastes that remain after spent nuclear fuel is reprocessed.

The technology involved in the acquisition of solar energy represents an interesting contrast with the technology of fossil and nuclear fuels. To begin with, no mining, refining or transport is involved, since solar energy is delivered—as sunshine—everywhere on the earth. If solar energy is to be used to provide heat, rather simple technology is called for, requiring only a solar collector (basically water-conducting pipes attached to a blackened metal plate in a glass-covered, weather-proof box) and conventional heat transfer equipment. If electricity is to be produced from solar energy, the technology is more complex, but the basic principles are well known, and in use. (Solar cells have reliably supplied electricity to satellites.) The technological issues that remain in this case are solely related to the problem of producing solar cells at a competitive cost.

3. The Efficiency of Energy Conversion.   As already indicated, the social value of energy is to be measured not by the amount of fuel consumed, but by the amount of useful energy released and, ultimately,
by the value of the resultant goods or services. Hence efficiency—the
amount of final goods or services yielded per unit of fuel-contained
energy employed—becomes an important consideration. In part this
involves issues that relate to the manner in which energy is used;
these are taken up in Vol. III. Here we consider the problem of maximizing the efficiencies with which power is produced from the original
fuel.

One approach is to reduce the number of steps in the process of
power production by combining stages, or through direct conversion of
chemical, nuclear or heat energy to electricity. One of these new technologies is the gas turbine. Here, the first two stages—heat accumulation and conversion of the heat to mechanical energy—are combined in
one apparatus. The gas turbine itself is not very efficient unless it is
operated at extremely high inlet temperatures, but before such high
temperatures can be maintained, the proper materials and effective
methods of cooling turbine blades must be found. It is possible, however, to gain an improvement over current efficiencies by combining the
gas turbine in combination with steam turbines and using the heat of
the gas turbine to power the steam boilers.

Two other methods aim to reduce the number of energy production
stages by direct conversion. The process of magnetohydrodynamics
(MHD) produces electricity by passing hot ionized gasses through a
strong magnetic field, thereby converting heat directly to electricity.
Development of practical MHD technology requires the materials that
can withstand the extreme speeds at which the gasses must travel,
and the extreme temperature differential created by heating the gasses
inside a duct while supercooling the magnets on the outside. Another
approach to the simplification of energy transfer systems is the fuel
cell which converts chemical energy directly to electricity. Fuel cells
(Chapter 5) will run on a variety of fuels, among them hydrogen,
natural gas, and carbon monoxide in combination with air or oxygen.
Since fuel cells can be designed to emit low levels of thermal waste,
they require little cooling water, emit no uncontrollable pollutants, and
thus are one of the most promising new technologies. Fuel cells large
enought to serve as auxiliaries to conventional power plants are now
in construction.

In considering the social value of different means of producing
energy, which is an essential step in developing a national energy policy,

careful consideration must be given to the interlocking effects of re-newability, the complexity of conversion technology, and its efficiency. Clearly an energy system that relies chiefly on non-renewable fuels (as the present U.S. system) is only a temporary expedient. Moreover, because the supply of such fuels is limited, increasingly intense efforts must be used to find and mine them, thereby requiring a rising level of capital investment—a social cost of increasing importance in present economic conditions. Finally, if the social value of energy is to be measured by the goods and services it can produce, then the efficiency with which a given source of energy can be converted into a useful form becomes a decisive factor in such a judgment. In reading the material presented in this volume, it may be useful to consider it in the context of these interactions. Also bear in mind that most of these articles were written in 1972. Although there have been some new technological developments since then, I feel that the general information contained herein is still very valuable.

October 2, 1974   **Barry Commoner**

ROBERT WILLIAMS

Fossil Fuel
Resources[1]

Much of the concern about fossil fuel resources stems from shortages of certain fossil fuels in recent years—in particular, natural gas has not been available for some new markets and coal supplies (especially low sulfur coal) have not been available to meet unexpected electrical utility demands. These problems, however, concern short term economic conditions and current regulatory policies more than any fundamental condition relating to the extent of energy resources (*Resources,* 1971). Besides concern about these immediate shortage problems, there have been perennial cries that we have fossil fuel reserves adequate to meet our needs for only a few more years. As we point out below, such statements about the inadequacy of reserves often reflect an especially restrictive definition of the concept "reserves." In discussing estimates of resources to be recovered in the future, it is important to keep in mind resource quantity concepts as well as the particular numbers involved. It is certainly true that our fossil fuel resources are limited. The critical questions are how long can we and should we rely on fossil fuels, what alternative energy conversion technologies are likely to be available before we exhaust our fossil fuel resources, and what are the various environmental and social trade-offs associated with these new technologies.

[1]Since this paper was written in 1971 a number of other important studies relating to fossil fuel resources have been completed. The interested reader may wish to examine the following, which are especially noteworthy: "USGS Released Revised U. S. Oil and Gas Resource Estimates," News Release, U. S. Dept. of Interior, March 26, 1974; "*U. S. Energy Resources, A Review as of 1972,*" a background paper prepared for the Committee on Interior and Affairs, U. S. Senate, Washington, D.C., 1974; "Mineral Resources and the Environment," National Academy of Sciences, Washington, D.C., February 1975.

1

The concern about the environmental impact of fossil fuel consumption is relatively new for society as a whole. However, such interest is not likely to be a fad. Rather if present trends continue, environmental concerns are likely to become more intense, owing to the exponential nature of the growth in both energy production and pollution burdens.

FOSSIL FUEL RESOURCE AVAILABILITY AND SOME LIMITS TO ITS UTILIZATION

The question of the availability of fossil fuel resources is one about which there is much disagreement—largely because different experts discuss quite different quantities.

To avoid confusion which might arise from uncertainties in basic terminology, clear definitions must be given to the most common terms used in making resource estimates: "reserves," "ultimately recoverable resources," and the "resource base." The term reserves refers to the stocks or inventory of the resource whose location is definitely known and which can be profitably extracted immediately or in the near future with current techniques. Not included in reserve estimates are known quantities regarded as too costly to produce profitably. Reserves constitute but a limited subset of the resource base, which is the total amount of the resource material within a given geographic area. The resource base includes not only reserves but also the rest of the resource material which can reasonably be presumed to exist, whether the deposits have been discovered or are still undiscovered, regardless of cost considerations or technical feasibility of extraction. In considering the future source of supply of a resource the term "reserves" is much too restrictive a concept; further exploration and the changes in technology and cost criteria that are almost sure to occur will make larger quantities available. It is equally certain that the "resource base" concept is much too inclusive; there is no reason to believe that anything like all of a resource presumed to exist in the environment can ever be exploited, even if it can ever be found. In the middle ground between "reserves" and the "resource base" is the useful concept of "ultimately recoverable resources," which is that part of the resource base which is judged to be technically and economically recoverable at some future time. Obviously the numbers that would be assigned to all three of these concepts are subject to constant change: (1) reserve estimates will change as the producer adjusts his inventory, (2) resource base estimates will change as our knowledge of the occurrence of the resource in the earth improves, and (3) estimates of ultimately recoverable resources will change both as our knowledge of the resource base improves and as our judgments of technical and economic recoverability change.

While fossil fuel resource estimates have increased as time has passed, there are fundamental physical limits to the amount of fossil fuel resources that man should utilize as an energy source. To get an idea of the fundamental limits involved in the problem, it is worth comparing the total amount of reduced carbon produced by photosynthesis and believed laid down in fossil deposits with various estimates of the resource base for fossil fuels:

Fossilized reduced carbon, formed photosynthetically from CO_2	68×10^{20} grams (Rubey, 1951)
Estimated total amount of carbon in shale containing more than 5% organic matter[2]	7.00×10^{20} grams (Duncan and Swanson, 1965)
Estimated total amount of carbon in world[2] coal deposits	0.12×10^{20} (Averitt, 1969)
Estimated total amount of carbon in world[2] petroleum deposits	0.01×10^{20} (Hendricks, 1965)
Estimated total amount of carbon in world gas deposits	0.005×10^{20} (Hendricks, 1965)

A striking feature of these estimates is that the carbon deposited in oil shale far exceeds that estimated for the other fossil fuels. Moreover, the sum of the fossil fuel resource estimates amounts to only about 10% of the total amount of carbon fixed in photosynthesis and believed to have been fossilized. However, even if fossil fuel resource estimates should be increased in the future, as more knowledge is obtained, the amount that can be utilized by man is limited by geophysical factors.

One limit on the amount of fossilized carbon that man can utilize as fuel is set by the availability of atmospheric oxygen. Burning only about 6% of the total amount of fossilized carbon (4×10^{20} grams) would deplete the atmosphere of its oxygen supply. But as we shall see this number is still large compared to the total amount of carbon contained in ultimately recoverable resources, believed to be about 8.2 $\times 10^{18}$ grams.[3] Should these resources eventually be burned, atmospheric oxygen would be reduced by about 2%. A more serious problem might be the effect of the increased atmospheric CO_2. The build-up of CO_2 can lead to a heating of the planet and climatic change through the

[2]The organic matter is assumed to be 75% carbon.
[3]On the basis of data compiled by Duncan and Swanson (1965), Averitt (1969), and Hendricks (1965), this total can be distributed among the various fossil fuels as follows (see the discussion in *Total Fossil Fuel Resources of the World*, below):

Oil shale	1.70×18^{18} grams
Coal	6.00
Petroleum	0.26
Natural Gas	0.24

so-called greenhouse effect, where the outgoing terrestrial radiation is trapped by CO_2 in the atmosphere. Burning all of the world's ultimately recoverable resources would release twelve times as much CO_2 as the present atmospheric loading. Over a period of centuries, much of this would enter the oceans—to what extent one cannot say for sure. At present about half of the CO_2 being released in the combustion of fossil fuels is being taken up by the biosphere and by the oceans. Over a long period of time such uptake could be at a much higher rate.

Both the build-up of CO_2 and the partial reduction of atmospheric oxygen provide limits to man's use of fossil fuels that are more significant than resource availability. Thus even if 8.2×10^{18} grams of fossil carbon is a low estimate of what could be ultimately recoverable, these geophysical factors would weigh heavily against burning more fossil fuel than this amount.

We now take a closer look at energy resources by fuel, first for the U.S. and then for the world as a whole.

TOTAL COAL RESOURCES OF THE U.S.

Coal-bearing rocks are widely distributed and abundant in most parts of the United States. Coal beds characteristically have great lateral continuity and geological mapping makes it possible to project them long distances from their outcrops with fair reliability. An extensive survey of coal resources of the U.S. has been prepared by Paul Averitt (1969) of the U.S. Geological Survey, and is widely regarded as the most comprehensive and authoritative work in the field. Averitt estimates that the total remaining U.S. coal resources of all ranks is over 3.2 trillion tons, of which about half may be considered recoverable. To get a feeling for the significance of this estimate, this number should be compared to other more familiar numbers. Total coal consumption in the U.S. in 1970 was 527 million tons, (U.S. Bureau of Mines, 1971). Thus, if about half of the 3.2 trillion tons is recoverable, our coal resources could meet our current level of demand for coal for 3000 years.

Table 1 summarizes much coal resource information according to rank, depth of deposits, location, and according to whether or not the various estimates are due to mapping and exploration. As the information here is quite concentrated it is fruitful to isolate certain portions of it for the sake of discussion.

The geographical distribution of coal resources is very unequal, as indicated in Figure 1. The bulk of the higher rank coals (anthracite and bituminous coal) lie primarily in the East, while lower ranked coals (subbituminous coals and lignite) are concentrated in the Northern

Great Plains and the West, including Alaska. These abundant lower rank coals are little used today, primarily because the rail transport capacity to the major coal markets is limited. However, low-rank coals are well suited for the production of synthetic gas and liquid fuels, and in many parts of the West they can be mined efficiently by stripping methods.[4] With these advantages low-rank coals of the West are certain to receive attention in the future (Averitt, 1969). The bulk of our coal resource is in categories favorable for mining; about half of our resource is bituminous or higher rank coal, about half lies in beds thicker than 28 inches, and most of it is buried less than 1000 feet.

In Table 1 it is indicated that about half of the total resources are in the category of estimated additional resources in unmapped and unexplored areas (largely resources in coal bearing areas mapped or examined only in reconnaissance). That 89% of the mapped and explored resources lie at depths less than 1000 feet merely reflects the fact that most exploration and mapping have been done only at shallow depths. Little is known about coal resources below 3000 feet.

It was stated above that about 50% of our coal resources can be regarded as recoverable. This means that past experience suggests that this percentage can be mined at or near present costs, measured in man-hours and equipment. Although the average recoverability of coal in strip mining is about 80%, strippable coal constitutes such a small fraction of the total resources (see below) that the overall recoverability factor of 50% is appropriate.

Surface Coal Resources

Averitt has estimated that the total remaining recoverable resources of surface coal in the U.S. to depths up to 150 feet are 128 billion tons, (assuming a recoverability of 80%), as of January 1, 1970 (Averitt, 1970). For comparative purposes this much coal:

1) Represents 8% of total recoverable coal resources.
2) Could supply the current level of coal consumption for 240 years.
3) Is 29 times the cumulative strip coal production as of January 1, 1970.

More recently the U.S. Bureau of Mines (1971) has indicated that about 46 billion tons of surface coal is recoverable *under present economic conditions*—less than 3% of all recoverable coal.

---

[4]However, a serious concern is whether stripped lands in arid parts of the West could be reclaimed to any useful function. If not, it would be unwise to develop these resources. See for example, National Academy of Sciences, *Rehabilitation Potential of Western Coal Lands*, a report to the Energy Policy Project of the Ford Foundation, Ballinger Publishing Co., Cambridge, 1974.

Table 1 Total Estimated Remaining Coal Resources of the United States, January 1, 1967 (Averitt, 1969, p. 12–13)

(In millions of short tons. Figures are for resources in the ground; about half of which may be considered recoverable. Includes beds of bituminous coal and anthracite 14 in. or more thick and beds of subbituminous coal and lignite 2.5 ft or more thick)

State	Overburden 0-3,000 ft Thick							Overburden 3,000-6,000 ft Thick	Estimated Total Remaining Resources in the Ground, 0-6,000 ft Overburden
	Resources Determined by Mapping and Exploration					Estimated Additional Resources in Unmapped and Unexplored Areas[a]	Estimated Total Remaining Resources in the Ground	Estimated Resources in Deeper Structural Basins[a]	
	Bituminous Coal	Subbituminous Coal	Lignite	Anthracite and Semi-Anthracite	Total				
Alabama	13,518	0	20	0	13,538	20,000	33,538	6,000	39,538
Alaska	19,415	110,674	b	c	130,089	130,000	260,089	5,000	265,089
Arkansas	1,640	0	350	430	2,420	4,000	6,420	0	6,420
Colorado	62,389	18,248	0	78	80,715	146,000	226,715	145,000	371,715
Georgia	18	0	0	0	18	60	78	0	78
Illinois	139,756	0	0	0	139,756	100,000	239,756	0	239,756
Indiana	34,779	0	0	0	34,779	22,000	56,779	0	56,779
Iowa	6,519	0	d	0	6,519	14,000	20,519	0	20,519
Kansas	18,686	0	d	0	18,686	4,000	22,686	0	22,686
Kentucky	65,952	0	0	0	65,952	52,000	117,952	0	117,952
Maryland	1,172	0	0	0	1,172	400	1,572	0	1,572
Michigan	205	0	0	0	205	500	705	0	705
Missouri	23,359	0	0	0	23,359	0	23,359	0	23,359
Montana	2,299	131,877	87,525	0	221,701	157,000	378,701	0	378,701
New Mexico	10,760	50,715	0	4	61,479	27,000	88,479	21,000	109,479
North Carolina	110	0	0	0	110	20	130	5	135
North Dakota	0	0	350,680	0	350,680	180,000	520,680	0	530,680

Table 1 (Continued)

State	Resources Determined by Mapping and Exploration (Overburden 0–3,000 ft Thick)					Estimated Additional Resources in Unmapped and Unexplored Areas[a]	Estimated Total Remaining Resources in the Ground	Overburden 3,000–6,000 ft Thick: Estimated Resources in Deeper Structural Basins[a]	Estimated Total Remaining Resources in the Ground, 0–6,000 ft Overburden
	Bituminous Coal	Subbituminous Coal	Lignite	Anthracite and Semi-Anthracite	Total				
Ohio	41,864	0	0	0	41,864	2,000	43,864	0	43,864
Oklahoma	3,299	0	d	0	3,299	20,000	23,299	10,000	33,299
Oregon	48	284	0	0	332	100	432	0	432
Pennsylvania	57,533	0	0	12,117	69,650	10,000[e]	79,650	0	79,650
South Dakota	0	0	2,031	0	2,031	1,000	3,031	0	3,031
Tennessee	2,652	0	0	0	2,652	2,000	4,652	0	4,652
Texas	6,048	0	6,878	0	12,926	14,000	26,920	0	26,926
Utah	32,100	150	0	0	32,250	48,000	80,250	35,000	115,250
Virginia	9,710	0	0	335	10,045	3,000	13,045	100	13,145
Washington	1,867	4,194	117	5	6,183	30,000	36,183	15,000	51,183
West Virginia	102,034	0	b	0	102,034	0	102,034	0	102,034
Wyoming	12,699	108,011	b	0	120,710	325,000	445,710	100,000	545,710
Other States	618[f]	4,057[g]	46[h]	0	4,721	1,000	5,721	0	5,721
Total	671,049	428,210	447,647	12,969	1,559,875	1,313,080	2,872,955	337,105	3,210,060

[a] Estimates by H. M. Beikman (Washington), H. L. Berryhill, Jr. (Virginia and Wyoming), R. A. Brant (Ohio and North Dakota), W. C. Culbertson (Alabama), K. J. Englund (Kentucky), B. R. Haley (Arkansas), E. R. Landis (Colorado and Iowa), E. T. Luther (Tennessee), R. S. Mason (Oregon), F. C. Peterson (Kaiparowits Plateau, Utah), J. A. Simon (Illinois), J. V. A. Trumbull (Oklahoma), C. E. Wier (Indiana), and the author for the remaining States.

[b] Small resources and production of lignite included under subbituminous coal.

[c] Small resources of anthracite in the Bering River field believed to be too badly crushed and faulted to be economically recoverable.

[d] Small resources of lignite in beds generally less than 30 in. thick.

[e] After Ashley.

[f] Arizona, California, Idaho, Nebraska, and Nevada.

[g] Arizona, California, and Idaho.

[h] California, Idaho, Louisiana, Mississippi, and Nevada.

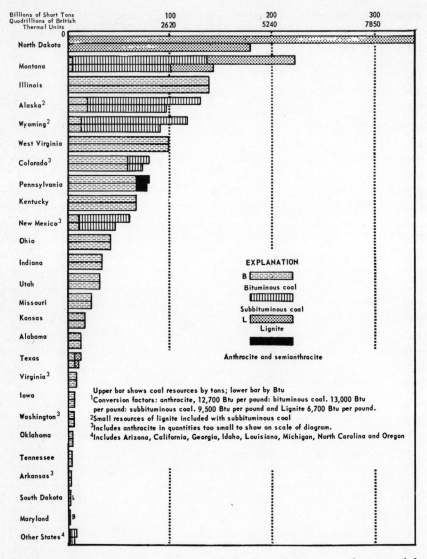

Figure 1 Remaining coal resources of the United States as determined by mapping and exploration, January 1, 1967, by States, according to tonnage and heat value. (Averitt, 1969, p. 20.)

The centers of the most intensive strip mining operations in the U.S. are the hilly northern Appalachian basin and the more flat Illinois basin. However, the largest concentration of strippable coal resources is the northern Great Plains region (western North Dakota, eastern Montana, and northeastern Wyoming), where 40% of the strippable resources are located. Although this coal is easily recoverable, relatively little mining is carried on in the northern Great Plains today, owing to the remoteness from major coal markets.

Low Sulfur Coal Resources

Public concern about the degradation of the air we breathe from the increasing burden of SO_2 produced in the burning of fossil fuels has led to restrictions on sulfur content of fossil fuels burned in many communities. In these restrictions the maximum allowable content is typically 1%, although some regulations call for a sulfur content less than 0.3%; New Jersey in particular issued rules calling for levels of 0.2% in coal after October 1971 (Squires, 1970). Accordingly, the demand for low sulfur coal has increased dramatically in recent years. This new demand, mainly on the part of the electrical utilities, is in addition to the traditional demand of steel producers for low sulfur coal to be used in the manufacture of coke. Unfortunately, adequate low sulfur coal resources are not readily available to meet these increased demands. This assertion can be demonstrated with statistics compiled in a 1966 Bureau of Mines survey of sulfur content in U.S. coals (De Carlo, 1966). In this report coal reserves were classified arbitrarily according to sulfur content as low sulfur (1% or less sulfur), medium sulfur (1.1% to 3.0% sulfur), and high sulfur (3.1% or more). Especially noteworthy are the following statistics:

1) About 65% of our coal reserves have low sulfur content.
2) However, 80% of the low sulfur coal is low rank lignite or sub-bituminous coal, most of which lies west of the Mississippi River remote from major coal markets.
3) Less than 10% of all low sulfur deposits (95 billion tons) lies east of the Mississippi River.
4) About half of the low sulfur coal reserves east of the Mississippi are concentrated in West Virginia.

Some of these relationships are illustrated in Figures 2 and 3.

Since only a small portion of our low sulfur coal reserves occur in the eastern part of the U.S., where the demand is greatest, it is clear that in the long run air quality standards will be met in the East by means other than by burning low sulfur coal in conventional coal-fired plants.

TOTAL OIL RESOURCES OF THE UNITED STATES

Much misunderstanding of the nature and magnitude of the country's oil resources has been the result of misjudgments rooted in ill-placed reliance on "reserves" statistics. For example, the chief geologist of the

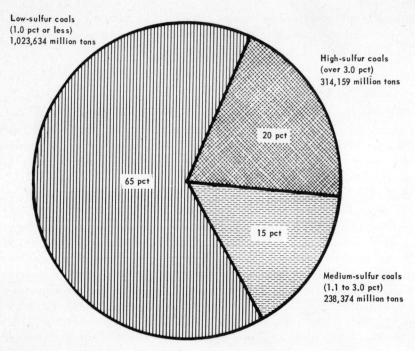

Low-sulfur coals
(1.0 pct or less)
1,023,634 million tons

High-sulfur coals
(over 3.0 pct)
314,159 million tons

20 pct

65 pct

15 pct

Medium-sulfur coals
(1.1 to 3.0 pct)
238,374 million tons

Figure 2 Estimated remaining coal reserves of all ranks, by sulfur content, in the United States, Jan. 1, 1965.

U.S. Geological Survey reported in 1920 that petroleum still in the ground and recoverable by contemporary methods amounted to no more than 7 billion barrels. It was highly improbable, he added, that the error in this estimate exceeded 50%. He felt that oil resources would be exhausted by 1934. In fact when 1934 came 12 billion barrels had been recovered and another 12 billion barrels had been determined as "proved reserves" (Landsberg and Schurr, 1968).

Petroleum reserve estimates are compiled and published annually by the American Petroleum Institute (API), the industry's top trade association. The estimates represent a select committee's best estimate of the volume of petroleum in known fields that can be profitably recovered with present technology.

While reserves data usually indicate only a few years advance supply of petroleum, these figures in themselves should not be a cause for alarm regarding the question of the adequacy of petroleum supplies to meet future demands. However, this does not imply that petroleum supplies are not in short supply. They are. But to see this, considerations of resources other than reserves must be made. To determine the adequacy of future supplies the resource base must first be examined.

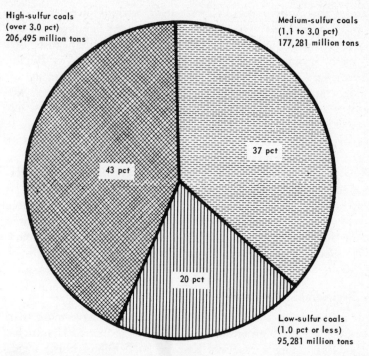

High-sulfur coals
(over 3.0 pct)
206,495 million tons

Medium-sulfur coals
(1.1 to 3.0 pct)
177,281 million tons

37 pct

43 pct

20 pct

Low-sulfur coals
(1.0 pct or less)
95,281 million tons

Figure 3 Estimated remaining coal reserves of all ranks, by sulfur content, in states east of the Mississippi River, Jan. 1, 1965.

The Resource Base

Estimates of the petroleum resource base (i.e., total oil originally in place) are few, and those that have been made are crude. A lower limit is provided by the total amount of oil discovered so far. Two estimates of the oil discovered so far are the following (in billion of barrels):

API (as of 12/31/69)   *Hendricks (1965)*
            395   400

About 93 billion barrels, or 23% of this, has been produced so far.

Where authorities disagree is on what fraction of the total resource base is represented by past discoveries and how the resource base relates to the amount of ultimately recoverable resources.

In a 1968 revision (United States Petroleum through 1980) of a 1965 study by T. A. Hendricks, it was estimated that there were originally 2000 billion barrels of crude oil in the ground in the U.S. (including the continental shelf out to a depth of 600 feet). This number was determined on the basis of geological comparisons between unexplored areas and known petroleum bearing regions. However, the estimation procedure is not very involved—in fact the simplicity of the procedure is an indication of how little is really known about how much oil remains in the ground. On the basis of information provided by the National Petroleum Council, Hendricks estimated that $2\frac{1}{2}$ million square miles of surface area of the U.S. could be regarded as favorable to the occurrence of petroleum. To locate all the petroleum would require exploratory wells with a spacing of one well every two square miles. If the average depth were 7000 feet, this would give 8.75 billion feet of exploratory drilling. So far, one billion feet of drilling has resulted in the discovery of 400 billion barrels, so that if future drilling activity is on the average half as successful, the total to be discovered should be 2000 billion barrels. Hendricks regarded his estimate of the resource base as conservative, since experience has shown that newly explored rock has time after time proved to be more productive than predicted. However, Hubbert (1969) regards Hendricks' estimate as optimistic.

Ultimately Recoverable Resources

Estimates of utimately recoverable petroleum resources have more than a two-fold range. At the high end of the estimates is that of Hendricks, whose method (assuming a resource base of 2000 billion bbls.) yields a total recoverable resource of 460 billion bbls. A low estimate of 190 billion bbls. has been provided by Hubbert, who gives no consideration to the resource base whatsoever. Rather Hubbert examines past drilling experience to arrive at his estimate.

Hendricks assumed that only $\frac{1}{3}$ of the rock potentially favorable for petroleum will actually be explored by drilling, and that the recovery rate for this exploration should average $\frac{3}{4}$ what it has been in the past. He also assumed that those resources which today are *technologically* recoverable, will someday be *economically* recoverable, which is about 40% of the amount discovered. On the basis of these assumptions Hendricks arrived at his estimate of 460 billion barrels.

In Hubbert's calculation of ultimately recoverable petroleum the trend in past discovery rates (barrels of petroleum per foot drilled) is projected into the future, without considering any geological factors.

Both of these estimates are very crude. A more recent (1970) independent estimate by the National Petroleum Council based on proprietary industrial data gives 432 billion barrels, just 6% below the Hendricks estimate. Since industrial estimates tend to be on the conservative side, one might argue that the Hubbert estimate appears low.

While the Hendricks and Hubbert estimates for the ultimately recoverable resources of petroleum are certainly crude, they are the best estimates available. It is interesting to note that if the demand for petroleum continues to increase at 4% annually from the present level of 5.4 billion barrels per year, Hubbert's estimate of remaining recoverable resources (190 − 93 = 97 billion bbls.) would be adequate to meet this demand for only 14 years, while the Hendrick's estimate (460 − 93 = 367 billion bbls.) would be adequate for 33 years. Neither figure offers any room for comfort. Even if consumption were stabilized at present levels the resource picture does not look too bright. In this case Hubbert's estimate would be adequate for

$$\frac{97 \text{ billion bbls.}}{5.4 \text{ billion bbls per year}} = 18 \text{ years}$$

while Hendrick's estimate gives

$$\frac{367}{5.4} = 68 \text{ years.}$$

Environmental considerations are likely to result in many more constraints on drilling activity than has been the case in the past. That this will be so has already been indicated in the broad-based opposition to the Trans-Alaska Pipeline, which resulted in a temporary postponement of its construction. As North Slope oil is developed, the country will lose at the very least one of its last remaining wildernesses, whether pipeline breakage should occur or not. The prospect of an earthquake-triggered pipeline rupture, spilling massive quantities of oil into the delicate arctic environment is not unlikely, either. A thorough discussion of the thermal effects of a heated pipeline in permafrost can be found in Lachenbruch (1970). In years to come we should expect to see more ecological disasters at sea such as accompanied the wreckage of the Torrey Canyon. Likewise, accidents in off-shore drilling, most notably in the Santa Barbara Channel, have resulted in ecological disasters which have raised public furor nationwide.

Oil exploratory activity is turning more and more to environments that pose more technological problems for the producer. The fact that these same environments appear to be more susceptible to ecological damage probably means that substantial constraints will be imposed on producers as a limiting factor in determining future recoverability.

Because of resource limitations, difficult future recovery conditions, and likely environmental constraints, the prospects for crude oil remaining a major energy source over the long term are not good. However, it must be stressed that there is much uncertainty as to the extent of our petroleum resources.

TOTAL NATURAL GAS RESOURCES OF THE UNITED STATES

From an environmental viewpoint, natural gas is an attractive fuel. Its combustion produces mainly water vapor and carbon dioxide, no fly ash, and little objectionable sulfur dioxide. Accordingly, natural gas is in high demand. But estimates of recoverable natural gas left in the ground indicate that natural gas is not likely to remain a significant resource much beyond the turn of the century.

The most limited concept used to describe gas supply is that of proved reserves, which refers to the stocks or inventory of the natural gas, whose location is definitely known and which can be profitably extracted immediately or in the near future with current techniques. Proved reserves constitute a limited subset of remaining recoverable resources. As in the case of crude oil the concept has always enjoyed far greater prominence than the limitations of the concept warrant. Proved reserve statistics are compiled and published annually by the industry-based American Gas Association.

The Resource Base

One of the few estimates available for the resource base of natural gas is that made by T. A. Hendricks of the U.S. Geological Survey in 1965, and subsequently revised in 1968. The Hendricks estimate for the natural gas resource base is directly related to his estimate for the petroleum resource base for the United States.

Hendricks noted that natural gas usually is associated with crude oil. He determined the ratio of occurrence from American Petroleum Institute and American Gas Association statistics for recovered oil and gas: 6182 cubic feet of natural gas per barrel of oil for 1952–1956, and 6637 cubic feet for 1957–1962. The rising trend in this ratio has resulted in part from the increasing effort to find and develop gas reserves and in part from the greater average depth of drilling. The ratio tends to be larger at greater depths. Hendricks assumed that in the future, the ratio would be 6667 cubic feet per barrel for economically recoverable quantities. Since the current recovery rate for crude oil is 30%, whereas for the more mobile natural gas it is 80%, the above ratio corresponds to a ratio of $\dfrac{6667/.8}{1/.3} = 2500$ cubic feet of natural gas occurring per barrel of crude oil originally in place. Thus the Hendricks' estimate of the resource base is

(2500 cf per barrel) $\times$ (2000 billion barrels) = 5000 trillion cubic feet.

A problem with this estimate is that natural gas is not always associated with petroleum. In the early days, natural gas was a by-product of crude oil exploration. At present natural gas exploration is often carried out independently of oil exploration. In fact, about 75% of the proved reserves and 75% of annual production of natural gas are for gas that is not associated with crude oil (AGA et al., 1970).

Nevertheless, this is the best estimate available for the resource base. Most knowledgeable geologists who could provide resource base estimates are employed by oil and gas producing companies, interested primarily in locating and controlling sufficient supplies to ensure operational continuity for a limited number of years.

It is of interest to compare the heat content of the resource base for each of the three major fossil fuel energy resources. The average heat contents of natural gas, oil, and coal are assumed to be the following:

10^3 BTU/cf of natural gas
6×10^6 BTU/barrel of crude oil
20×10^6 BTU/ton of coal.

The best available estimate for the coal resource base is 3.2 trillion tons, as provided by Paul Averitt (1969) of the U.S. Geological Survey. The heat value for the resource base of each of these fuels is thus:

Natural gas	5×10^{18} BTU
Crude oil	12×10^{18} BTU
Coal	64×10^{18} BTU

Thus in terms of what is in the ground, there is four times as much coal as natural gas and crude oil combined, on an equivalent heat value basis, according to the best available information. As discussed below, estimates of what is ultimately recoverable favor coal as the more abundant resource. The heat value of the sum of these three fossil fuel resources is more than a thousand times the total U.S. energy consumption in 1970, whereas the natural gas resource base is only 72 times annual consumption.

Ultimately Recoverable Resources

Two noteworthy and different estimates of ultimately recoverable gas resources are provided by T. A. Hendricks (1965) and by M. K. Hubbert (1969). Another estimate of ultimately recoverable resources, defined somewhat more restrictively than in the case of the first two, was provided biennially for several years by the Potential Gas Committee (1971).

(a) The Hendricks estimate of ultimately recoverable resources of natural gas is related to his estimate of the amount of crude oil that will be discovered eventually, which is 1150 billion bbls. Multiplying this figure by the ratio of occurrence (2500 cubic feet [cf] of natural gas per barrel of crude oil originally in place) and by the 80% recoverability factor gives 0.80 × (2500 cf/barrel) × (1150 billion barrels) = 2300 trillion cf of natural gas that should eventually be recovered. Thus Hendricks estimated that slightly less than half of the amount he estimated for the resource base would be recoverable ultimately.

The main flaw in this estimate is that it is based on the assumption that oil and gas are associated. Moreover, future drilling is going to be primarily in deeper wells, where even associated gas should have a higher occurrence ratio than past experience suggests.

(b) The Hubbert estimate for ultimately recoverable natural gas resources, makes no consideration of the resource base (and hence geological evidence). Rather he based his estimate entirely on trends in past drilling experience. As in the case of the Hendricks estimate, he assumed that gas recovered in the future will be associated with crude oil, but in the ratio 7500 cf of natural gas per barrel of crude oil (higher than the Hendricks figure). Hubbert estimated that the ultimately recoverable resources of petroleum would be 190 billion barrels, of which only 29 billion barrels would correspond to discoveries in new fields, so that the portion of ultimately recoverable gas resources from discoveries in new fields would be (29 billion barrels) × (7500 cf/barrels) = 218 trillion cf. In addition he estimated that additional gas would be found in fields already discovered totaling 222 trillion cf. The proved discoveries (i.e., proved reserves plus cumulative production) as of January 1, 1967, totaled 604 trillion cf. Thus his total estimate is:

Proved discoveries as of 1/1/67	604
Additional gas in already discovered fields	222
Future discoveries	218
Separate estimate for Alaska	150
Total	1194 trillion cf

The Hubbert estimate is thus about half the Hendricks estimate.

Both the Hubbert and the Hendricks estimate suffer from the questionable assumption that oil and gas occur in a fixed ratio. Moreover, the Hubbert estimate is limited by its lack of consideration of geological evidence.

(c) The Potential Gas Committee, made up of representatives of various segments of the natural gas industry has provided estimates of the potential supply of gas that can be expected to be found in the future under these assumed conditions:

1) Adequate but reasonable prices.
2) Normal improvements in technology.

This potential gas supply is in addition to the proved reserves estimated by the American Gas Association. Three categories of the potential gas supply (probable, possible, and speculative) are given. Estimates made for 1966, 1968, and 1970 are as follows:

	As of 12/31/66	As of 12/31/68	As of 12/31/70
Probable	300 trillion cf	260 trillion cf	257 trillion cf
Possible	210	335	387
Speculative	180	632	534
Total	690 trillion cf	1227 trillion cf	1178 trillion cf

The significant increase in the estimate from 1966 to 1968 is due mainly to the inclusion of over 400 trillion cf potential for Alaska. Also the offshore depth limit was increased from 600 to 1500 feet and drilling depths were increased from 25,000 feet to 30,000 feet.

In order to compare the potential gas supply estimates with the ultimately recoverable resource estimates of Hendricks and Hubbert, cumulative production and proved reserves figures must be added. According to the American Gas Association, cumulative production as of December 31, 1970, was 392 trillion cf, while reserves totaled 291 trillion cf. Thus the amount of ultimately recoverable resource would be:

$$1178 + 392 + 291 = 1861 \text{ trillion cubic feet.}$$

This estimate is about half-way between the Hubbert and the Hendricks estimates.

Although the Potential Gas Supply estimate does not allow for any significant changes in gas recovery technology and, being an industry-based committee, is likely to have judged future economics conservatively, it is probably the best available estimate of what could be recovered with economic and technological conditions characteristic of the time the estimates were made.

As was true for the resource base, it is of interest to compare the heat values of the fraction of total remaining fossil fuel resources which are regarded as ultimately recoverable. We take the Hendricks (1965) estimates for both crude oil and gas. To get remaining recoverable resources, cumulative production must be subtracted, giving 2300–392, or about 1900 trillion cf of natural gas, and 460–93, or about 370 billion barrels of oil. We use the Averitt estimate for coal. (Averitt regards coal resources as 50% recoverable. Cumulative production of coal has been insignificant.) We obtain:

	Natural Units	Heat Content
Natural gas	1900 trillion cf	1.9×10^{18} BTU
Crude oil	370 billion bbls.	2.2×10^{18} BTU
Coal	1.6 trillion tons	32×10^{18} BTU

There is thus more nearly eight times as much energy in remaining recoverable coal resources as in gas and oil combined. It should be noted that natural gas and crude oil are about equally scarce. The remaining recoverable natural gas resource amounts to 85 times current production and would correspond to a 44-year supply if demand grows at 3.3% per year (corresponding to the mean annual rate of growth of total energy since WW II). If the more rapid natural gas growth rate of 7.2% were sustained, this supply would last only 28 years. We have already seen that if oil use continues to grow at 4% per year the resource is estimated to last 33 years.

It is clear that if estimates for ultimately recoverable crude oil and natural gas resources are at all meaningful, then these resources are in quite limited supply and cannot be expected to retain their present significances as sources of energy beyond the turn of the century.

Supplemental Gas

Several estimates of the amount of natural gas that can be recovered in the next 10–15 years indicate that natural gas supplies may fall short of demand. While such forecasts usually involve rather questionable assumptions about trends, such predictions have stimulated interest in substitutes for domestically produced natural gas. Whether supply falls short of demand in the next 10 years or in the next 25, such substitutes ought to be investigated fully.

Canadian Natural Gas Imports.   Canadian gas resources are abundant. The resource base is estimated to be at least as large as that of the "lower 48" states of the United States. With energy use in Canada only about $\frac{1}{12}$ that in the U.S., exports to the U.S. (currently about 42% of production) may increase somewhat in the future. The National Energy Board of Canada estimated natural gas surpluses available for future export, taking into account both expected rates of discovery, and probable future Canadian needs. For 1975 this surplus was estimated to be 1.9 trillion cubic feet, while the projection for 1980 is a lower 1.1 trillion cubic feet (Smith, 1971). Additional gas exports hinge upon the increase of the Canadian discovery rate beyond the historical average for the last ten years. There are indications that such an increase may take place in the Canadian Arctic, an area exploitable through pipelines.

Liquified Natural Gas. At atmospheric pressure, natural gas has a boiling point of minus 259°F. Since the liquid occupies 1/632 of the corresponding gas volume, natural gas transport in the liquid state is becoming attractive for shipping gas long distances. World-wide tanker commerce in liquified natural gas (LNG) is expected to develop in the seventies. The problem of catastrophic systems failures has not been documented.

Nuclear Stimulation of Gas Reservoirs. It has been estimated that there is about 317 trillion cubic feet of gas resources (in addition to the potential gas supply estimated by the Potential Gas Committee) in tight gas-bearing formations that could be made available through "stimulation" by nuclear explosives. The idea of "nuclear stimulation" came out of the Atomic Energy Commission's Plowshare Program, which seeks to find peaceful applications of nuclear explosives. In tight gas-bearing formations, natural gas is present in the pores of the rock, but the pores are very poorly connected. By setting off a nuclear device in this rock a fracture system is set up connecting many of these pores. It is hoped that the effective diameter of the well can thereby be increased from the bore diameter of six inches up to several hundred feet.

Environmental concerns play an important role in nuclear stimulation activities. The problems fall in two general areas—radioactivity effects and seismic effects. A nuclear explosive results in the creation of large quantities of deadly radioactive fission products, which can enter the biosphere in one of three ways: 1) venting of radioactive material at the detonation point, either at the time of explosion or later, 2) contamination of ground water that may be transported from the site and consumed directly or introduced into the food chain, 3) contamination of the gas product. In regard to the third problem, both radioactive krypton-85 and tritium are part of the gas product. Tritium can actually replace ordinary hydrogen in the methane molecule (principal natural gas constituent) and thereby become chemically fixed in the natural gas, with no ready means of removal. So as to minimize the tritium production, an all fission explosive is desirable (there is much more tritium produced when a fusion device is exploded). Curiously, in discussing nuclear weapons, an all fission device is regarded as "dirty," whereas a fusion device is regarded as "clean," since the radioactive fallout from the latter has less deadly fission products per kiloton. In gas stimulation, the reverse terminology is used, since an all fission device "dirties" the gas less. Whereas there may be no significant venting or ground water contamination in some explosions, seismic damage to surface and subsurface structures cannot be avoided.

If initial experiments at nuclear stimulation of natural gas are regarded as successful, plans call for setting off thousands of shots in various parts of the Rocky Mountains in Colorado and Wyoming in coming years.

The second nuclear gas stimulation experiment was Project Rulison, in which a 40-kiloton nuclear explosive was set off 8440 feet below the earth's surface near Grand Valley in western Colorado on September 10, 1969. If Rulison is regarded as a success, then 100 shots of 200 kilotons each would be set off over a 10-year period (one a month) in order to "develop" the Rulison field of 36,000 acres. This development would leave behind the fission products of 20 megatons of fission devices—a thousand times as much as the fission products produced by the Hiroshima bomb. Radioactivity from the Rulison shot was effectively contained. It is highly questionable whether adequate assurances could be given for the containment of all such future shots. Even if all the shots could be successfully contained, other resources in the ground become contaminated, precluding their retrieval in the foreseeable future. Moreover, it is unlikely that the local population would tolerate repeated seismic shocks. The seismic motion from the single Rulison underground blast was more than some local residents had anticipated, or were led to believe. Some 260 claims totaling more than $72,400 for structural damage had been paid as of April 1970 (Haas, 1971).

It is questionable whether Plowshare devices can offer an acceptable means of recovering natural gas resources. Little attention has been given to alternative means of stimulating gas production in tight reservoirs. The possibility also exists of stimulating gas production using conventional explosives. In this case there is no radioactivity. This alternative ought to be examined closely.

Synthetic Gas. The technical feasibility of producing pipeline quality gas from coal and lignite has been demonstrated, although the economics of gassification are uncertain. Coal supplies are abundant, although their recovery from underground and surface mines have produced serious environmental problems.

The details of coal gasification are discussed elsewhere. Here it will merely be mentioned that in 1969 the FPC predicted that by 1980 commercial scale coal gasification plants producing synthetic pipeline gas could provide from 231 to 500 billion BTU/day.

Gas from Organic Wastes. Another potential substitute for natural gas is methane produced from organic wastes. The potential from only such localized sources of animal manure as feedlots, dairy operations, and turkey and chicken operations amounts to about 3 trillion cubic feet of methane per year—nearly 80% of residential gas space heating needs.

TOTAL FOSSIL FUEL RESOURCES FOR THE WORLD

Remaining recoverable resource estimates for the world can be compiled from the works of Averitt (1969) for coal, Hendricks (1965) for oil and natural gas, and Duncan and Swanson (1965) for oil shale.[5] The definitions of ultimately recoverable coal for the U.S. are adopted for the world as a whole. For example, ultimately recoverable bituminous coal is taken to be 50% of coal in place in seams thicker than 14 inches with less than 6000 feet overburden. Ultimately recoverable petroleum is taken to be 40% of that to be discovered, while the corresponding figure for natural gas is taken to be 80%. In estimating ultimately recoverable resources of oil shale it is conservatively assumed that only high grade oil shale (yield greater than 25 gallons of oil per ton of shale) will be eventually recoverable. Some deposits with lower grade shale (as low as 10 gallons per ton) have actually been processed commercially in some parts of the world. Inclusion of these resources would increase the resource estimate by an order of magnitude.

Ultimately recoverable fossil fuel resources, in common energy units, are shown in Figure 4. Coal is relatively abundant, followed by oil

[5]Also Canadian tar sands figures from 1969 estimates by the Canadian Petroleum Association are given. Tar sands resource estimates are not readily available for the rest of the world.

Figure 4 World-remaining recoverable resources

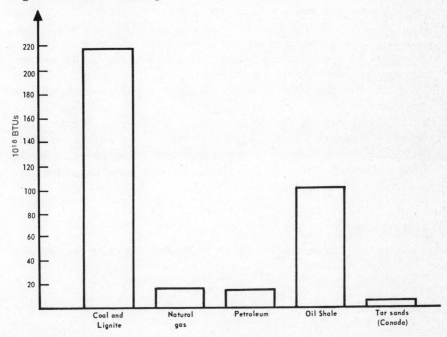

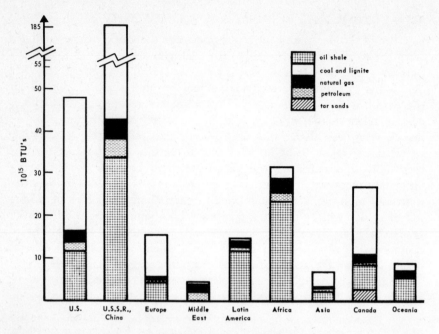

Figure 5   Remaining recoverable resources by region

shale. Natural gas and petroleum are in relatively short supply. Thus, the high rate of consumption of these high grade fuels cannot be sustained. Within a couple of decades, these fuels will have to be supplemented by synthetic fuels from coal, oil shale, tar sands, or even from organic wastes, or by hydrogen produced at nuclear plants. If the 10 gallon per ton level is used, oil shale greatly exceeds coal in abundance.

The distribution of remaining recoverable resources in nine world regions is shown in Figure 5. More than 50% of the world's remaining recoverable fossil fuels are in the USSR and China (most lie in the USSR). The U.S. takes second place. Note the potential importance of oil shale in certain parts of the world.

REFERENCES

American Gas Association; The American Petroleum Institute; the Canadian Petroleum Association, 1970. Reserves of crude oil, natural gas liquids, and natural gas in the United States and Canada and United States productive capacity as of Dec. 31, 1969. Vol. 24.

Averitt, Paul, 1969. *Coal resources of the United States, January 1, 1967.* U.S. Geological Survey Bulletin 1275.

————, 1970. *Stripping coal resources of the United States, January 1, 1970.* U.S. Geological Survey Bulletin 1322.

Brooks, David B., and John V. Krutilla, 1969. *Peaceful uses of nuclear explosives: Some economic aspects.* Resources for the Future Publication.

Bureau of Natural Gas, 1969. *A Staff Report on National Gas Supply and Demand.* Federal Power Commission, Washington, D.C.

DeCarlo, J. A., et al., 1966. *Sulfur content of United States coals.* U.S. Bureau of Mines Information Circular 8312.

Duncan, D. C., and V. E. Swanson, 1965. *Organic-rich shale of the United States and world land areas.* U.S. Geological Survey Circular 523.

Haas, Paul, 1971. The Rulison project in retrospect. *Nuclear News,* May.

Hendricks, T. A., 1965. *Resources of oil, gas, and natural gas liquids in the U.S. and the world.* U.S. Geological Survey Circular 522.

————, 1968. *United States petroleum through 1980.* U.S. Department of the Interior.

Howell, W. D., and J. B. Hille, 1970. Explosive detonation tested in hydraulically fractured gas wells. *J. Petrol. Tech.,* Apr.

Hubbert, M. K., 1969. Energy resources. In Freeman, *Resources and man.* Committee on Resources and Man, National Academy of Sciences, National Research Council.

Lachenbruch, Arthur H., 1970. *Some estimates of the thermal effects of a heated pipeline in permafrost.* Geological Survey Circular 632.

Landsberg, H. H., and S. H. Schurr, 1968. Energy in the United States: Sources, uses, and policy issues. In *Resources for the future.*

National Petroleum Council, 1970. Future petroleum provinces of the United States.

Potential Gas Committee, 1971. Potential supply of natural gas in the U.S. as of Dec. 31, 1970.

Resources No. 36, 1971. *Behind the energy crisis.* Resources for the Future Publication.

Rubey, W. W., 1951. *Geological history of sea water.* Bulletin of the Geological Society of America.

Squires, Arthur M., 1970. Clean power from coal. *Science* 169 (Aug. 28): 821.

U.S. Bureau of Mines, 1971. U.S. Energy use sets new record. News Release, Mar. 9, 1971.

U.S. Bureau of Mines, 1971. *Strippable reserves of bituminous coal and lignite in the United States.* U.S. Bureau of Mines Information Circular.

Werth, Glenn C., et al., 1971. An analysis of nuclear explosive gas stimulation and the program required for its development. UCLRL-50966, Livermore, Calif., Lawrence Radiation Laboratory.

ROBERT H. WILLIAMS

Fission Fuel Resources[1]

Of all the nuclear isotopes occurring naturally in practical concentrations, only the uranium isotope ^{235}U fissions (i.e., it splits up into lighter atoms) when bombarded by slow neutrons. When a uranium atom fissions a portion of its mass is converted into energy, roughly one hundred million times the energy released in the chemical combustion of a hydrocarbon molecule of a fossil fuel. On a mass basis the fissioning of 1 gram of ^{235}U releases, on the average, 8.19×10^{10} joules (0.945 megawatt days) which is equivalent to the heat of combustion of 3 tons of coal or 14 barrels of oil.

Most naturally occurring uranium is not the fissionable isotope ^{235}U, however. In fact only 0.71% of natural uranium is ^{235}U; almost all the remainder is the isotope ^{238}U. Although ^{238}U is itself not readily fissionable, neutron bombardment of this isotope leads to the formation of a fissionable isotope of the element plutonium, ^{239}Pu. Because it can be converted into a fissionable material, ^{238}U is called a fertile isotope. A second substance that has this property is thorium, whose isotope ^{232}Th can be converted to the fissionable ^{233}U.

Power generation based on the fission process must use ^{235}U as initial fuel. In the current generation of reactors little use is made of the primary constituent of natural uranium, ^{238}U. In fact with present large light water reactors (LWRs) only about 0.71 grams of fissionable

[1]Since this paper was written in 1971 a number of other important studies relating to fission fuel resources have been completed. The interested reader may wish to examine the following, which are especially noteworthy: P. K. Theobold et al., "Energy Resources of the United States," U. S. Geological Survey Circular 650, Washington, D.C., 1972; Thomas B. Cochran, The Liquid Metal Fast Breeder Reactor, An Environmental and Economic Critique, Resources for the Future, Inc., Washington, D.C., 1974.

plutonium are produced for every gram of ^{235}U that is consumed. Since these reactors consume more fuel than they produce they are called burners. The next generation of reactors will be capable of utilizing much more of the ^{238}U in uranium ore. Since in fact they will produce more fissionable material than they consume they are called breeders. Large commercial breeder reactors, however, are not expected to go into operation at least until sometime in the 1980s, so that in the meantime the present generation of burner reactors will be highly dependent on our ^{235}U reserves.

UNITED STATES URANIUM RESOURCES

Uranium, one of the less common elements, makes up about two parts per million (ppm) of the earth's crust (Singleton, 1968), with traces found nearly everywhere. Granite, the principal parent rock of the continents, contains 4 ppm of uranium, although the Conway Granite of New Hampshire contains about 12 ppm.

Major U.S. ore deposits, containing between 2000 and 3200 ppm of uranium, are located in the western part of the country. About 95% of the ore occurs in flay-lying bedded deposits in fluvial sediments, mostly sandstone. The remaining 5% occurs in lignite and vein deposits (Bieniewski et al., 1971).

The principal low-grade sources of uranium are the phosphate rock deposits (100–200 ppm) in Florida, Montana, Idaho, Wyoming, and Utah; the Chattanooga Shale (70 ppm), primarily in Tennessee, and the large copper deposits (6 ppm) of the western United States.

Uranium resources are generally given as the tonnage of U_3O_8 that can be profitably recovered from the ground at a given price per pound. In 1968 the price paid for U_3O_8 by the AEC and private buyers averaged $7.57 per pound. Previously the price has been somewhat higher. During the period 1942–1970 the federal government acquired a total of 326,000 tons of U_3O_8, at an average cost of about $9 per pound. In the early years the U.S. was quite dependent on foreign sources of uranium, with 150,000 tons of federal purchases having come from abroad (USAEC, 1970). Today the U.S. has evolved into the world's leading uranium producer.

Uranium ores are presently being removed from deposits both by open-pit and underground mining methods. In 1968, 62% of the uranium recovered came from ores in underground mines, 36% from open-pit mines and 2% from miscellaneous sources (Bieniewski et al., 1971). The largest low cost uranium reserves are located in New Mexico and Wyoming, with each state accounting for about 40% of the reserves under $8 per lb. (as of January 1, 1970).

A potential low-cost (less than $8 per lb.) source of uranium is available as a by-product of copper mining operations in the western U.S., in copper leach solutions. It is estimated that about 1000 tons of U_3O_8 per year for 30 years would be recoverable from this source. Another potential source of U_3O_8, at about $10 per lb., is the 85,000 tons that should be available as a by-product in the so-called wet-process manufacture of phosphoric acid produced from phosphate rock. A vast low grade (70 ppm) source of uranium would be the 2.6 million tons of U_3O_8 available from the Chattanooga Shale, which underlies about 40,000 square miles in the southeastern U.S. and which is estimated recoverable at a selling price of about $69 per pound. An even vaster store of even lower grade (12 ppm) uranium is the Conway Granite of New Hampshire which is estimated to contain more than 3 million tons of U_3O_8 per hundred feet of depth. However, recovery costs would be in excess of $100 per lb. (Bieniewski et al., 1971).

One of the largest uranium mines is the sea, which contains uranium at a concentration of .003 ppm, nearly a thousand times less than the earth's crust. Although this "ultimate resource" of 4 billion tons is exceedingly dilute, British scientists have recovered gram quantities of uranium in a coastal laboratory-scale plant. Surprisingly a recent estimate of recovery costs is about $20 per lb. for the British technique of absorption on insoluble compounds of such elements as lead and chromium (O.E.C.D., 1970). Estimates of others range from $30 to $100 per lb. In making the lower estimates certain problems (e.g., what about fish?) inherent in processing enormous quantities of water were neglected. It is not likely that uranium will be recovered from seawater on a large scale in the near future (Bieniewski et al., 1971).

Several estimates of U.S. uranium resources in various price ranges are listed in Table 1. Higher priced ores are generally not tabulated because of inadequate information. The estimates are subdivided into two categories, "Reasonably Assured Resources" and "Estimated Additional Resources." The former refers to reserves in the usual mining sense, that is, to uranium which occurs in known ore deposits of such grade, quantity, and configuration that it can, within the given price range, be profitably recovered with currently proven mining and processing technology. The term Estimated Additional Resources refers to uranium surmised to occur in unexplored extensions of known deposits or in undiscovered deposits in known uranium-bearing districts, and which is expected to be discoverable and economically recoverable in the given price range. The reliability of resource estimates is of course greater for the reasonably assured resources than for the estimated additional resources. Likewise the figures are more reliable for the price range $5–$10 per pound than for higher priced ranges, since prospecting efforts have been primarily directed toward finding low-cost uranium. It is noteworthy that in the $5–$10 range the resource estimates have roughly doubled in the period 1966–1971. In addition to the quantities listed here it is expected that the AEC will release to the commercial market the equivalent of 50,000 tons of U_3O_8 from its production stockpile (O.E.C.D., 1970).

These various uranium resource estimates are best understood in relation to the fuel requirements of a nuclear power plant. A typical light water reactor capable of generating 1000 megawatts of electrical power (1000 MW(e)) consumes about 2½ kilograms of fissionable material in one day, if the plant operates 80% of the time. If the fuel is uranium enriched to 3.3% in ^{235}U, then about 1230 lbs. (more than half a ton) of U_3O_8 must be processed to provide fuel for one day's operation. The uranium ore requirements are of course much greater than this. A 1000 MW(e) plant would require over 200 tons of high grade ore (2500 ppm U_3O_8) per day. In contrast a comparable sized coal burning plant would need about 7000 tons of coal per day. Thus on a tonnage basis about 35 times as much coal must be mined to produce the same amount of electricity as a ton of uranium ore. For lower grade ores this mining advantage of uranium is diminished. For uranium mined from the Chattanooga Shale (70 ppm U_3O_8) the tonnages of coal and uranium ore needed to produce a given amount of electricity would be comparable. (It should be noted that the surface mining problems for uranium may become comparable to those for coal when low grade ores are recovered.)

How long will our low cost uranium resources last with present day reactors? Let us consider a very rapid growth scenario. In 1971 nuclear generating capacity amounted to 9200 MW(e). This capacity could grow to about 200,000 MW(e) by 1985 (corresponding to an average growth rate of 22% for 1971–1985) and to 1,200,000 MW(e) by 2000 (corresponding to a 12% growth rate for 1985–2000) if USAEC expectations are realized. If these plants use 80% of their capacity, on the average, then the total amount of uranium resources under $10/lb (1.07 million tons of U_3O_8) would be exhausted by about 1995. Thus low cost uranium resources appear to be in relatively short supply. The critical question of course is how much more costly uranium resources will effect the price of electric power, since there are increasing quantities of uranium at higher prices. We will return to this later.

WORLD URANIUM RESOURCES

The largest indicated world resources outside the U.S. (and excluding the Soviet Union, Eastern Europe, and Mainland China) lie in Canada

Table 1 Estimates of U.S. Uranium Resources (thousands of short tons of U_3O_8)

	(USAEC, 1962)	(USAEC, 1966)	(OECD, Dec. 1967)	(USAEC, Jan. 1969)	(USAEC, Jan. 1970)	(OECD, Apr. 1970)[b]	(USAEC, Jan. 1971)	(Bureau of Mines, 1971)
$5–$10 per lb.								
Reasonably Assured								
Conventional	175	200	300	200	250	250	300	206
By-Product				120	90	90	90	30
Estimated Additional								
Conventional	420	325	350	325	600	600	680	
By-Product				25				
Total	595	525	650	670	940	940	1070	
$10–$15 per lb.								
Reasonably Assured								
Conventional		150	150		140	140		100
By-Product					20			85
Estimated Additional								
Conventional		200	200		300	300		
By-Product					50			
Total		350	350		510	440		
$10–$30 per lb.[a]								
Reasonably Assured								
Conventional	400		350	200	280		340	200
By-Product				150	20		20	85
Estimated Additional								
Conventional	300		640	675	1000		980	
By-Product								
Total	700		990	1025	1300		1340	

[a]Includes $10–$15 per lb. estimates.
[b]It is expected that an additional 50,000 tons of U_3O_8 will be released from AEC stockpiles.

Table 2 Estimates of World Resources of Uranium
(thousands of short tons of U_3O_8)

	(OECD, 1967)	(OECD, 1970)	(USAEC, 1970)
$5–$10 per lb.			
Reasonably Assured	826	930	929
Estimated Additional	743	884	974
Total	1569	1814	1903
$10–$15 per lb.			
Reasonably Assured	770	750	786
Estimated Additional	556	661	711
Total	1326	1411	1497
$10–$30 per lb.			
Reasonably Assured	1358		
Estimated Additional	1960		
Total	3318		

and South Africa, where reserves under $10 per lb. total 2.32×10^5 tons and 2.0×10^5 tons of U_3O_8, respectively (Organization for Economic Cooperation and Development, 1970). World resources (including U.S. resources) are given in Table 2.

Demand for uranium ore concentrate in coming years is expected to put a severe strain on U.S. low cost uranium resources with the projected growth of nuclear power from light water reactors. The question naturally arises whether foreign sources can be exploited to meet domestic needs. There are hard economic reasons why this should be not expected. For one, although foreign resources are greater than those in the U.S. there are sharp limitations on foreign production capability— notably the South African production is tied to gold mining and the Canadian production is restricted to a limited number of mines. Foreign nuclear generating capacity is also rapidly growing. The OECD forecasts for 1980 an installed "free" world nuclear generating capacity of 300 thousand megawatts, as compared to about 145 thousand for the U.S. (O.E.C.D., 1970). Accordingly, it is expected that a world-wide seller's market will develop for uranium. In such a situation foreign utilities will probably be able to outbid U.S. utilities for foreign uranium. The reason for this is that in Europe and Japan fossil fuels are much more expensive than in the U.S.—averaging $5 to $10 per ton more (in terms of coal or coal equivalents) than U.S. utility fuel costs. In 1968 the average price of coal imported to Japan was about $16 per ton, and in 1967 Ruhr coal in West Germany cost $16–$17 per ton (U.S. Bureau of Mines, 1969). In contrast the average price of coal as burned by U.S. electric utilities in 1969 was $6.22 per ton (National Coal Association, 1970). This means that, all other factors being equal in the nuclear-vs-fossil competition, utilities in these areas will be able to pay much more for U_3O_8 than their U.S. counterparts and still maintain a competitive posture vis-a-vis fossil fueled plants.

THORIUM RESOURCES

Although there has been little demand for thorium to date, and hence little prospecting for it, the estimated U.S. thorium resources are substantial, as shown in Table 3. Since ^{232}Th is a fertile rather than a fissionable isotope, demand for thorium is likely to develop only if and when a breeder reactor program becomes well established. No significant commercial market appears likely in the foreseeable future. On a world-wide basis current estimates are that 500,000 tons of ThO_2 can be regarded as Reasonably Assured Resources under $10 per lb. and that about 1,000,000 tons can be considered Estimated Additional Resources (O.E.C.D., 1970).

In the near term the principal use of thorium as a nuclear fuel in the U.S. will be in two high temperature gas cooled reactors—one 30 MW(e) power station at Peach Bottom, Pennsylvania, which was started up in 1966, and the 330 MW(e) power station at Fort St. Vrain, Colorado. These advanced converter reactors use ^{232}Th as a fertile material along with uranium highly enriched to over 90% (Dukert, 1970) with fissionable ^{235}U, an enrichment level much higher than the 2–3% typical of present light water reactors.

Table 3 Estimates of U.S. Resources of Thorium
(USAEC, 1962) (thousands of short tons of ThO_2)

$5–$10 per lb.	
Reasonably Assured	100
Estimated Additional	300
Total	400
$10–$30 per lb.	
Reasonably Assured	100
Estimated Additional	100
Total	200
$30–$50 per lb.	
Reasonably Assured	3000
Estimated Additional	7000
Total	10000

FUEL CYCLE

Uranium fuel, unlike coal, requires considerable processing before it is ready for use in a reactor. The processing operations can be broadly categorized as milling, refining and conversion, enrichment, and fabrication. The uranium resource costs cited above are for the product of the milling operation only, namely for the uranium concentrate, known in the industry as "yellow cake," which usually assays between 70% and 90% U_3O_8.

Milling.   In the milling process uranium ore is first pulverized and is then contacted with either an acid or an alkaline carbonate which dissolves the uranium, a step known as leaching. The dissolved uranium is selectively recovered from the "leach liquor" and is calcined (roasted) to remove excess water. The product of this operation is yellow cake (Hogerton, 1964).

Refining and Conversion.   In refining uranium, crude concentrates (yellow cake) from uranium mills are purified by solvent extractions and then calcined to form essentially pure uranium trioxide (UO_3) known as "orange oxide." The orange oxide is first converted to uranium dioxide (UO_2) and then to uranium tetrafluoride (UF_4) called "green salt," by reaction with hydrogen fluoride gas. The green salt is then reacted with fluorine gas (F_2) to form the volatile uranium hexafluoride (UF_6) used in the enrichment process (Hogerton, 1964). (Although it is a solid at room temperature, UF_6 vaporizes at slightly elevated temperatures.)

Enrichment.   Light water reactors require enriched uranium as fuel. When ordinary water is used as the reactor moderator the uranium fuel must be enriched in the isotope ^{235}U to compensate for unwanted neutron absorption by hydrogen in the water. The U.S. nuclear power industry inherited enrichment services from the AEC's nuclear weapons program—probably a determining factor in this country's decision to develop light water reactors for the first generation of nuclear power. (Heavy water moderated reactors can use natural uranium as fuel.) The task of enriching uranium so that it contains (typically in LWRs) 2.2%–3.4% ^{235}U is the most difficult step in the chain of uranium fuel processing and merits special attention.

In the enrichment process a partial separation of the uranium isotopes is accomplished, resulting in a product, with a higher-than-normal concentration of ^{235}U, and a waste (or tailing) of depleted uranium, having a lower-than-normal concentration of that isotope. Because both isotopes ^{235}U, and ^{238}U are the same chemical element, separation cannot be achieved by ordinary chemical methods. Rather the methods used must be based on mass-dependent properties. Since the mass difference is small (235 vs. 238), rather sophisticated separation techniques are required.

The technique used in the U.S., known as the "gaseous diffusion" method, (see Von Halle et al. for a detailed discussion) makes use of the different rates of diffusion of gases through a porous barrier, the gases being $^{235}UF_6$ and $^{238}UF_6$. In a gas mixture of two isotopes the lighter molecules will diffuse more rapidly through a porous barrier than will the heavier species, resulting in a slight enrichment of the lighter isotope in the gas passing through the barrier. The degree of enrichment possible in diffusing through a single porous barrier, called the "separation factor," is the square root of the mass ratio, 1.0043 in this case—a number very close to 1.0. This means that many stages of separation are required to accomplish any significant degree of separation of the uranium isotopes. To produce 90% enriched uranium from natural uranium requires about 3000 separation stages in series.

In the enrichment process a large portion of the natural uranium used is discharged from the enrichment plant as "tails." For example, 6.07 kilograms of natural uranium are required to produce one kilogram of uranium enriched to 3.3% in ^{235}U. The remaining 5.07 kilograms is in the form of "tails" containing 0.2% ^{235}U. These tails are stored for use later as fertile material in breeder reactors.

Typically a gaseous diffusion plant consists of hundreds of stages of separation equipment connected in series in what is known as a diffusion "cascade." Because of the repetitive nature of the process—continuous diffusion through thousands of stages—the plants are of necessity among the largest industrial facilities in the world. In 1964 they consumed 45 billion kilowatt-hours of electricity (4% of the total electrical energy consumption in the U.S.) for driving the gas circulating pumps, etc. Altogether it is calculated that power consumption for enrichment will be 4.5% of the power generated in the nuclear plant (Gast, 1971). In addition the plants have a total cooling water circulation capacity of about 1.4 billion gallons per day (USAEC, 1968).

The U.S. currently has three large enrichment plants, originally built for defense purposes, and owned by and operated under contract with the AEC. Plants at Oak Ridge, Tennessee and Paducah, Kentucky, are operated by Union Carbide Corporation, while the plant at Portsmouth, Ohio is operated by the Goodyear Atomic Corporation. In spite of the immensity of these facilities, their capacity to meet enrichment needs, both domestic and foreign, is expected to saturate in the late 1970s.

The total present diffusion plant capacity could support the operation of about 131,000 MW(e) of power from LWRs (J.C.A.E., 1970). The U.S. is expected to have an installed nuclear generating capacity of nearly 150,000 MW(e) by 1980 and will provide additional enrichment services for about 90,000 MW(e) of foreign nuclear generating capacity by that date (some foreign contracts are for 30 years and can't legally be cancelled) so that significantly greater enrichment capacity is needed soon (Johnson, 1971). Demand will overtake capacity in about 1976. However, the advance inventory that will be available at that time and the additional production capacity that can be obtained by improvement of existing plants make it possible to postpone the need for additional plants until about 1980. New enrichment capacity is expensive. About $10 million is needed to install the capacity needed to provide enriched uranium for one 1000 MW(e) power plant.

Typically the operations of conversion and enrichment require a total of about 7 months for processing.

Fabrication.   The last step in the processing of reactor fuel is the conversion to the appropriate chemical form (UO_2 for light water reactors) and the fabrication of the fuel elements in the appropriate geometry and with the appropriate cladding.

The conversion step is a fairly straightforward chemical operation. Uranium dioxide is produced by reacting uranium hexafluoride first with water and then with an hydroxide salt. A precipitate results which is calcined to form orange oxide (UO_3), and this in turn is reduced with hydrogen to form UO_2 powder.

In fabricating the fuel elements the geometry is critical, since a fixed spatial distribution of fuel within the reactor core is required for the system to function properly. Also, because enormous quantities of heat are generated within a very small volume, it is essential to maintain proper channels for coolant flow through the core.

The fuel material is also enclosed in a thin protective sheath or "cladding" so as to protect the fuel from corrosion or erosion by the reactor coolant, to lock in the deadly radioactive fission products which are formed when fuel atoms undergo fission, and often to serve a structural function.

In the fabricating process small cylindrical pellets are compacted from UO_2 powder, and the pellets are then loaded into thin-walled cladding tubes, made either of stainless steel or an alloy of zirconium called "Zircaloy." An inert gas (helium) is then introduced into the tubes, and the tubes are end-capped. A cluster of such tubes separated by means of spacer devices then constitutes a fuel element (Hogerton, 1964).

The time required for the fabrication process is typically seven months. After fabrication the fuel is ready to be inserted into the reactor. However, after the fuel is delivered to the reactor site there is typically a two month delay before that fuel is inserted into the reactor and is contributing to the commercial operation of the plant (USAEC, 1970).

Disposal of Spent Fuel Rods

A very important part of the fuel cycle involves handling and reprocessing spent fuel rods when they are removed from the reactor, following burnup.

As the spent fuel rods are intensely radioactive, great care must be exercised in their removal from the reactor and in subsequent handling. Following removal from the reactor a fuel rod is first stored for about three months on site so as to allow some of the short-lived radioisotopes to decay prior to shipment to the chemical reprocessing plant (Povejsil, 1967). For transportation the spent fuel rods must be carefully loaded into enormous shielded casks. Spent fuel transport is estimated to require three months (Povejsil, 1967).

At the chemical reprocessing plant plutonium and uranium are separated from the fission products. It is noteworthy that in assessing the costs of reprocessing (which takes about two months) provision is made for storage in existing underground liquid waste disposal tanks, but it is customary to ignore the costs of reducing the wastes to solid form and the costs of perpetual storage for such solid wastes in underground vaults (e.g., in salt mines). Also no account is taken of the costs for safeguarding plutonium or highly enriched uranium against diversion to weapons use. In Wash, 1099, for example, it is stated (USAEC, 1971) that "No allowance has been made for the costs of inspection to safeguard against diversion of fissile material to unauthorized use, on the basis that this is a national or international policing cost not to be charged to the reprocessing plant and its customers."

Overview of the Fuel Cycle

The entire fuel cycle from mining through reprocessing is indicated schematically in Figure 1. The numbers indicated correspond to the production of one kg of uranium fuel enriched to 3.3% and "burned" for 36 months in a pressurized water reactor. The parameters characterizing the fuel burnup are shown in Figure 2.

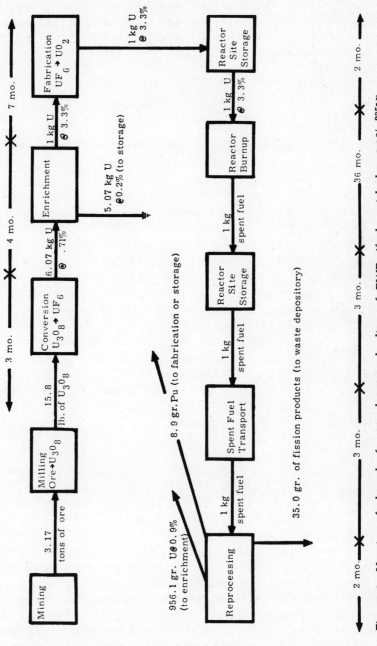

Figure 1 Uranium fuel cycle for steady state loading of PWR (fuel enriched to 3.3% ^{235}U)

36   **Robert H. Williams**

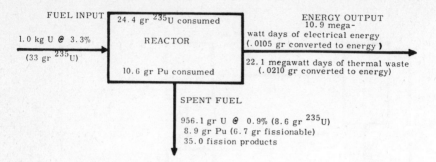

Figure 2   Typical PWR fuel burnup

It is interesting to note the rather large quantities of feed material necessary to produce one kg of fuel. When the tailings in the enrichment process are discharged at 0.2%, the feed material required is 6.07 kg of natural uranium, corresponding to 15.8 lb. of U_3O_8. A much larger quantity of uranium ore is required. When the uranium concentration in the ore is 2500 ppm, the ore required is 3.17 tons.

In Table 4, the charges and credits for a typical PWR fuel cycle are tabulated. It should be noted that the three most important costs are capital charges, enrichment, and mining and milling costs, in that order. The reason working capital charges are so high is that the fuel cycle is long (60 months here), with much time already spent before revenues are available from the generation of electricity.

Ultimately what is needed in fuel cost calculations is the fuel component of the cost of generating electricity, the sensitivity of this cost to various parameters, and a comparison to the corresponding cost for fossil fuel plants. (The appropriate cost for a coal-fired plant would be the "as burned" cost of coal.) In Table 5 costs for the uranium fuel cycle are compared to the 1969 average for the "as burned" costs of coal (National Coal Association, 1970). It is seen that for U_3O_8 at $8 per lb., the uranium fuel cycle costs are 1.96 mills/kwh, (without any charge for control and safety of fissile materials or for storage and disposal of radioactive wastes), as compared to 2.26 mills/kwh for coal. If capital costs are added to fuel costs the total cost of generating electrical power amounts to about 8 mills/kwh for both nuclear and fossil fuels. Doubling the price of uranium oxide increases the cost of generating electricity by only about 8% while doubling the price of coal leads to a 30% increase in the cost of generating electricity. Thus only a marked stabilization of coal prices should hurt the competitive position of nuclear plants relative to coal plants, unless capital costs (which are greater for nuclear plants) should substantially escalate or unless pres-

Table 4 Fuel Cycle Costs per Kilogram of Enriched Uranium for a PWR with Fuel Enriched to 3.3% ^{235}U

A. Charges

Fuel Cycle Step	Product	Output	Unit Cost	Step Cost
1. Mine and Mill	U_3O_8	15.8 lb. (6.07 kg U)	$8/lb.	$127.00
2. Conversion	UF_6	6.07 kg U	$2.74/kg U	16.60
3. Enrichment	enriched UF_6	1.0 kg U @ 3.3% 5.07 kg U @ 0.2%	$32/unit of separative work	159.00
4. Fabrication	fabricated UO_2	1.0 kg U	$90/kg U	90.00
5. Spent Fuel Transport			$5/kg of spent fuel	5.00
6. Reprocessing[a]	uranium, plutonium, fission products	956.1 gr U @ 0.9% 8.9 gr Pu 35 gr fission products	$33/kg of spent fuel	33.00
Working Capital Charges			12.6%[b]	168.10
Total Charges				$598.70

B. Credits

Recycled Quantity			Unit Credit	Total Credit
1. Fissionable Pu	6.7 gr fissionable Pu		$8/gr	$58.50
2. Uranium	956.1 grams @ 0.9%		$8/lb. plus separative work credits	27.20
Total Credits				$85.70

[a]Note that this does not include the cost of providing security for fissile by-products of the fuel cycle, or the costs of solidifying and storing in perpetuity the radioactive wastes from the spent fuel.

[b]This rate is the fixed charged rate on nondepreciable capital given on pages 3–11 of Wash 1098. It consists of an average earning rate of 7.0%, a federal income tax of 4.8%, and state taxes totaling 0.8%.

Table 5 Fuel Cycle Component of Electric Power Generating Costs for PWR

U_3O_8 Costs	$8 per lb.	$16 per lb.
Total Charges	$598.70	$807.70
Total Credits	$ 85.70	$105.70
Fuel Cycle Costs	5.13×10^5 mills	7.02×10^5 mills
Fuel Cost per kwh	1.96 mills/kwh	2.58 mills/kwh
Average As Burned Cost of Coal by Utilities[a] (1969)[20]	2.26 mills/kwh	2.26 mills/kwh

[a]Assuming a thermal efficiency of 40% for the power plant.

In the calculations leading to these figures 1 kg of U enriched to 3.3% was assumed to burn for 36 months (out of the total fuel cycle of 60 months), thereby generating 2.62×10^5 kwh of electrical energy.

ently ignored costs (such as those of safeguarding fissionable materials against diversion to weapons use and costs of perpetual storage of radioactive wastes) prove to be very high.[2]

Impact of the Fast Breeder on Uranium Resources

The nuclear fuel resource picture should actually improve if the breeder reactor becomes the dominant means of producing electric power, since the breeder actually produces more fissionable fuel than it consumes. Accordingly, the cost of electricity should become even more insensitive to the cost of natural uranium and lower grade ore deposits could be economically exploited. Should the breeder reactor program become well established there would be ample supplies of low grade uranium to last for tens of centuries in the Chattanooga shale of the Southeast, in the Conway granite of New Hampshire, or even from sea water.

Whether the fast breeder reactor becomes economically available in the 1980s or not, however, the uranium demand picture does not change appreciably in the 20th century. Cumulative uranium requirements are shown in Figure 3 for a nuclear power economy that grows to 735,000 MWe capacity by the year 2000. Even if the breeder reactor (with a 12-year plutonium inventory doubling time) is introduced at the unlikely early date of 1980, cumulative uranium requirements by the year 2000 are reduced only 24% from what they would be without the breeder. If the breeder is delayed 10 years—due either to unforeseen

[2]Since this was written capital costs for electric power generation have escalated greatly.

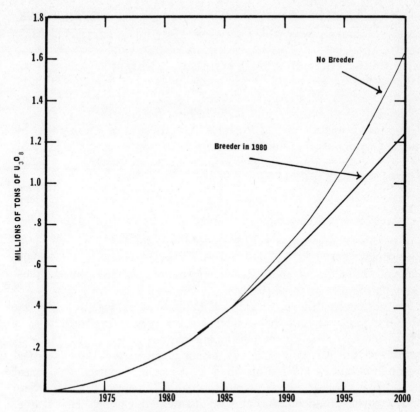

Figure 3   Cumulative U_3O_8 requirements for tails discharged at 0.2%

technological difficulties or serious environmental problems—the reduction in cumulative uranium demand below that anticipated without the fast breeder would be only 8%. In any case cumulative U_3O_8 requirements are not much affected by the breeder before the end of the century.

Comparing these demand projections with the resource estimates of Table 1 we see that with any high nuclear growth uranium prices may increase substantially before the year 2000. However, a doubling or even a quadrupling of the price of uranium should not appreciably affect the competitive position of nuclear power—even when heavy reliance on the light water reactor is maintained, since the cost of generating nuclear electricity is so insensitive to the price of uranium.

While fast breeder reactor economics are even more insensitive to fuel prices, it does not appear that fuel price considerations alone provide compeling motivation for the rapid development of the fast breeder reactor. Moreover, even with a relatively early introduction of the fast breeder, the cumulative uranium demand picture would not be appreciably affected until well into the 21st century, when it is likely that the fast breeder reactor would be competing with other new energy technologies such as solar energy or even controlled thermonuclear fusion.

Resource availability is not likely to be a limiting factor in the development of nuclear power. But other factors such as those posed by the need to maintain radioactive wastes in perpetual surveillance or by the risks of diversion of fissionable materials to weapons use are likely to pose problems with no easy answers.

Nuclear Power and Developing Nations

Nuclear fuels are compact and cheaply transported long distances. Accordingly, it might be thought that nuclear power could be the economic salvation of those developing nations lacking abundant indigenous supplies of fossil fuels.

Unfortunately this fuel transportation cost advantage is offset by other economic factors that render nuclear power unattractive as a major source of energy for developing nations. First of all, the economics of nuclear power favors existing large-scale interconnected power installations much more than in the case of power from fossil fuels. Thus, nuclear power is most suitable for those regions where electric power production has already attained a high level. Few, if any, electric power grids in less-developed nations are capable of absorbing nuclear reactors in the more economically favorable capacity range of 600–1000 megawatts. Also a high capital investment is required for a nuclear power plant (the capital costs are typically 25% higher than for a fossil fuel plant). Thus capital must be available at a relatively low interest rate for nuclear power to compete best. However, this is usually the opposite from the financial circumstances of the less-developed nations (except in those countries having abundant petroleum reserves, where nuclear power would not be presently competitive anyway). Thus nuclear power will be most attractive on purely economic grounds where fossil fuel costs are high, where a large electric power capacity already exists, and where the costs of borrowed money are low. Because of its scale and its capital intensity nuclear power will probably not help the power starved, less-developed countries very much in the foreseeable future. Nuclear power will most likely tend to widen rather than narrow economic and technological gaps between developed and developing nations. If economics were to guide the worldwide development of nuclear power then it is likely that its expansion would be limited mainly to the highly industrialized nations of the world.

REFERENCES

Averitt, Paul, 1969. *Coal resources of the United States, January 1, 1967.* U.S. Geological Survey Bulletin 1275.

Bieniewski, Carl, et al., 1971. Availability of uranium at various prices from resources in the United States. Bureau of Mines Information Circular 8501, U.S. Department of the Interior.

Dukert, Joseph, 1970. *Thorium and the third fuel.* Understanding the Atom Series, U.S.A.E.C.

Gast, Paul F., 1971. Power consumption and resources utilization in nuclear reactor fuel cycles. Memo to Gordon Goodman, June 7, 1971. Calculations based upon flow sheets in Wash-1099 (U.S.A.E.C. 1971).

Hogerton, John F., 1964. *Atomic fuel.* Understanding the Atom Series, U.S.A.E.C.

Joint Committee on Atomic Energy (J.C.A.E.), 1970. Uranium enrichment pricing criteria. Hearings, 91st Congress, 2nd Session, June 16 and 17.

————, 1971. Joint Committee on Atomic Energy announces hearings on uranium enrichment pricing. Press Release No. 650, Feb. 18.

Johnson, Wilfred E., 1971. The enrichment crisis: U.S. policy, requirements and capabilities. *Nuclear News,* p. 29, Jan. 1971.

National Coal Association. *Steam-electric plant factors/1970 edition.*

Organization for Economic Cooperation and Development (O.E.C.D.), 1970. European Nuclear Energy Agency and the International Atomic Energy Agency. *Uranium: Resources, production, and demand,* Sept. 1970.

Patterson, John A. Outlook for nuclear fuel, 1970. Presented at the IEEE-ASME Joint Power Generation Conference, Pittsburgh, Pennsylvania, Sept. 29, 1970.

Povejsil, D. J., et al. 1967. *Financial aspects of the nuclear fuel cycle.* Proceedings of the American Power Conference, vol. 29.

Singleton, Arthur L., Jr., 1968. *Sources of nuclear fuel.* Understanding the Atom Series, U.S.A.E.C.

U.S.A.E.C., 1967. *Forecast of growth of nuclear power.* Dec. 1967, Wash 1084.

————, 1968. *AEC gaseous diffusion plant operations.* ORO-658, Feb. 1968.

————, 1970a. *The Nuclear industry, 1970.*

————, 1970b. *United States nuclear industry fuel supply survey.* Division of Raw Materials, May 1970.

————, 1970c. *Potential nuclear power growth patterns.* December 1970. Wash 1098.

————, 1971a. *Forecast of growth of nuclear power.* Jan. 1971. Wash 1139, Division of Operations Analysis and Forecasting.

————, 1971b. *Reactor fuel cycle costs for nuclear power evaluation.* Wash 1099. USAEC Division of Reactor Development and Technology, Mar. 1971.

————, 1971c. News Release, May 12, 1971.

U.S. Bureau of Mines. *Minerals Yearbook, 1969.*

U.S. Bureau of Mines News Release, Mar. 9, 1971. U.S. Energy Use Sets New Record. U.S. Department of Interior.

Vernon, J. M., 1971. *Public investment planning in civilian nuclear power.* Duke University Press.

Von Halle, E., et al., 1965. Diffusion separation methods. In *Encylopedia of chemical technology.* 2nd ed., Vol. 7.

ARTHUR M. SQUIRES

The Fossil Fuel Development Gap[1]

Our fossil fuel industries lack technologies to meet urgent needs. Yet their own research establishments are not grappling with their problems in timely fashion. Why are these industries in such disarray? Why are they not doing more to find means for meeting future needs?

It will be useful to address these questions before asking, what is to be done? What are the most promising paths of development?

The picture is not entirely dark. A bright spot is the ongoing effort to build better gas turbines, funded at hundreds of millions each year, largely in response to desire for better aircraft. Commercial gas turbines will be available within a decade which will permit the designer of stationary power plant to achieve an efficiency of conversion of fuel to electricity on the order of 50%. The certain prospect of these machines creates an imperative for at least one major development in fossil fuel technology: means for supplying clean fuel to the gas turbines. This line of development, fortunately, will create opportunities for other fossil fuel industries as well—pipeline gas, liquid fuel, steel, and petrochemicals.

The current mood of executives of larger electricity companies is another positive factor. They understand that a process trial, to be meaningful, must be big in scale. In meeting their responsibilities to the nuclear program, they learned how to acquire new expertise quickly, and they came to appreciate the role of R&D in the introduction of a

[1]Work at The City College of New York on cleaning hot power gas is supported by Research Grants AP-00692 and AP-00945 from the Environmental Protection Agency.

Mr. Seymour B. Alpert of Stanford Research Institute, Menlo Park, California 94025, assisted generously with information and advice.

43

radical change in procedure. Discussions in the industry now point to a large increase in outlays for R&D, probably supported by an assessment against electricity sales. Recently, these executives have shown their readiness to invest huge sums in tests of equipment to suppress emissions of sulfur dioxide. Unfortunately, as we shall see, they may not get full value for investments made to date. Their mood augurs well, however, for speedy attack upon finding ways to furnish clean fuel to gas turbines.

R&D POSTURE OF THE OIL INDUSTRY

The central problem of the oil industry is future supply. Every expert predicts a sharp rise during the 1970s in imports of oil. It was a surprise, therefore, when news came that two major companies had closed down laboratories for research in synthesizing oil from coal and oil shale (Atlantic-Richfield at Dallas and Humble at Baton Rouge). The closing of Shell's laboratory at Emoryville, California, long famed as a center for fundamental studies in petroleum science, was also a shock. M. W. Kellogg's laboratory at Piscataway, New Jersey, although not as productive as this company's earlier facility in Jersey City, on which the oil industry used to rely so heavily, is also gone.

Why is process development for the oil industry contracting? Why is the industry so unconcerned over meeting the gap in oil supply, which otherwise can be met only at expense of our Nation's trade balance?

One reason may relate to the uncertainty surrounding government regulations for the quantity and price of imported oil. Development is hard to plan if the reward from the development can be altered drastically by the stroke of a bureaucrat's pen. In this connection, it may be remarked that oil industry research has not contributed notably to providing answers to the problems of the regulated fossil fuel industries, natural gas and electricity.

A more important reason, in our opinion, is the present scale of operations in the oil industry. Process development for the industry is costly and risky. Oil executives may well be making the correct choices viewed from their stockholder's interests, if not from National interest. Alpert (1971) has provided an example illustrating costs and risks. He listed nine efforts to develop processes for desulferizing residual oil "directly." Alpert guesses that each of the nine developers has invested from $5 to $20 million in R&D, including the cost of demonstration units. There is risk of outright failure: Alpert notes that three large units representing two processes have been shut down.

There is also risk of lack of acceptance, especially in view of the number of teams in the contest. There is even the possibility that *none* of the processes will achieve wide commercialization, since each can handle only selected feedstocks to achieve sulfur levels of about 1% rather than the desired level of 0.3%. We will have more to say on this point later on.

Alpert believes this may be the last competitive "race" to develop an operation treating oil at a large throughput. If so, it is the end of an era in petroleum R&D. The oil industry's efforts in the 1930s to find equipment of ever-increasing capacity, yielding superior products at lower cost and higher efficiency, may fairly be said to have set down the foundations of chemical engineering as a profession. The technique of fluidizing solids to bring them into intimate contact with gases was a direct outgrowth of these efforts, for example. With passage of time, oil R&D became more concerned with better catalysts, a goal achievable at moderate costs, than with development of new means for handling materials undergoing treatment, a goal achievable only at ever-increasing costs as unit processing capacity advanced. In 1930, the first unit of a new oil-conversion process might handle only a few thousand barrels per day, yet make money if successful, and not cost too much if a flop. Today, the first profit-making unit must handle scores of thousands of barrels per day, and cannot prudently be built until an expensive but unprofitable unit of far smaller size has been run. The last major American innovation in materials processing is the "ebullated bed," a conception of the mid-1950s on which millions had been spent by the early 1960s. Yet this has not yet achieved commercial acceptance, and was recently set back by a process failure which must have broadcast waves of caution through the minds of petroleum executives who plan R&D.

Architect-engineering firms catering to the oil industry have changed character. Successful firms have little in-house R&D. Even M. W. Kellogg, whose R&D was so vital to the oil industry of the 1930s, has adopted the pattern of its major competitors. Universal Oil Products notably remains active in development, but has long concerned itself primarily with goals which depend upon finding new catalysts rather than new means for handling materials.

Processing of coal and oil shale to provide synthetic fuels, gaseous or liquid, must inevitably begin with a major materials-processing step. The same may be said for processing the "dregs" from the refining of petroleum, to which little attention has been paid by oil company R&D.

The first vital step in which the raw material is attacked will most often determine the character of the remaining steps of the process.

The first step will often determine the nature of the products and probably even the overall efficiency.

It does not appear that we can look to petroleum company R&D for improved approaches to the materials-processing step, let alone for novel approaches. It is not that skills are lacking. On the contrary, the expertise available in petroleum companies and in architect-engineering firms catering to this industry must be viewed as a major asset to the Nation in setting up a program to tackle the problems of all the fossil fuel industries. Ingenuity will be needed to find the best way to exploit this asset.

R&D POSTURE OF THE GAS INDUSTRY

The central problem of the pipeline gas industry is also one of supply, and is much more urgent than the problem of the oil industry. For the latter, so long as the industry can expect import allowances to be advanced along with our increasing appetite for liquid fuels, there will be little incentive to shift to alternatives of high capital cost based upon domestic raw materials. For the gas industry, there is no low-cost alternative to domestic gas. Importation of liquefied natural gas is capital intensive. So also is manufacture of Synthetic Natural Gas (SNG) from cheap fuels such as coal and residual oil. Manufacture of SNG from distillate petroleum feedstocks requires less capital, but the feedstocks are dear. Alternatives to domestic gas all appear to cost more than 80¢ per million British Thermal Units (BTUs). Most alternatives cost more than $1. This must be compared with the latest well-head price of 26¢.

The industry long ago recognized that new supplies of gas would be needed shortly after 1970 (see Elliott, 1961, for example). Predictions of gas demand had to be revised sharply upward each year through the 1960s. Why is the pipeline gas industry so little prepared to meet its problems with results of its own R&D?

The industry was founded on plenty. In the mid-1940s, gas was being flared in Texas and elsewhere, and the industry began under a sense of urgency to move the gas and put it to use. The industry did not depend upon R&D for its beginnings (like oil, power, steel, and indeed most modern industry). Its executives did not think much about R&D, for this activity had little to do with next year's profits. The industry's laboratory at The Institute of Gas Technology was as much concerned with gas utilization as with alternative sources of supply, and in any case it has always had a modest budget. In spite of much good work, it does not have processes ready to meet the industry's need for new gas.

The industry is now building a number of plants to manufacture SNG from distillate petroleum feedstocks. It is ironic that these plants will use processes of British, Japanese, and German origin, for which license fees will be paid abroad. It will be instructive to review briefly the origins of these processes.

The British Gas Council's laboratory at Solihull, which opened in 1951, inherited a vigorous British tradition for research on better ways to make town gas from coal. Experiments on the hydrogenation of coal at high temperature and pressure, with object of obtaining the highest possible yield of methane, were conducted as early as 1936. The original intention at Solihull was to build and operate large-scale demonstration plants gasifying coal. The Solihull team built several such plants and made major contributions to our knowledge of coal gasification. In 1956, the emphasis shifted to making gas from petroleum feedstocks. The Solihull team had already sharpened its skills on large plants handling coal, and it was able quickly to develop two successful processes, one for gas from light naptha, and a second for gas from gas oil. A favorable factor was the relatively small scale of many of the Gas Council's individual plants, making it profitable to introduce a new process in equipment of relatively low capacity. The first plant of Solihull's design went into operation in 1964. By now, Solihull has accounted for more than 80 commercial installations and has earned more than 1,000,000 pounds in royalties. In 1972, Solihull is a going concern devoting its energies to vigorous defensive development toward maintaining its lead. Its potential earnings for the British balance of payments may be judged from Alpert's (1971) estimate that satisfaction of one-third of our own anticipated short-fall of gas in 1980 by processes based upon petroleum feedstocks would mean payment of $100 to $200 million in royalties to foreign licensors of the required process technology.

Japan appears to have a process for gas from light naptha which may be an improvement upon the British; it claims ability to use a naptha of wider boiling range and higher sulfur content. The Japanese advantage, if real, appears to rest upon an improvement in catalyst formulation. Alpert (1971) writes that Japan has apparently got a leg up on the rest of the world in catalyst development. Germany also has a process for gas from light naptha.

American research is not likely to find a process challenging the Japanese, British, and German technologies and also completely free from patent infringement. Solihull has a process for gas from crude oil, but the process is not applicable to all crudes. This, or a process to make gas from residual oil, would appear to be a better target for American effort.

Solihull's director, F. J. Dent, one of the great development engineers of his generation, wrote upon his retirement in 1966: "The cost of operating on . . . (Solihull's) scale calls for a fairy godmother and ours has been the Gas Council. Budgets have increased more than 20 times while they have been in control." The Gas Council got its money's worth.

Dent (1966) also wrote: "It is significant too, that we usually had reason to regret any protracted period of exploratory laboratory investigation. Small-scale experiments have often been time-wasting even when large-scale conditions have been reproduced as faithfully as possible. . . . Operation on a reasonable scale at an early stage is most desirable to throw difficulties into their proper perspective. Laboratory work was of most value after the problems had been recognized in this way."

In a process development where the central problem is one of processing a large amount of liquid or solid in an initial step (as was the case for Solihull in its work on gas oil and crude oil), money is often wasted if budgets are too small to allow an early, direct attack upon this problem on the practical scale.

In a nutshell, this is the explanation for the tardiness of American efforts to develop processes to provide SNG from coal. The budgets of the Office of Coal Research and Bureau of Mines, even when supplemented by industry contributions, have simply been too small. Experimentation on the first coal-processing step on a practical scale is just now getting underway for one of the processes under study, is a few months to a year away for two others, and is in the planning stage for a fourth. At present level of funding, each process is at least ten years away from operation of a first full-scale plant, and 20 years away from making a significant contribution to our gas supplies. Can we afford to wait so long?

Except for a small, sustained effort by the Bureau of Mines, one promising candidate for the first coal-processing step in a plant for SNG from coal has been neglected, viz., treating raw coal with hydrogen at high pressure and high temperature. Data in the literature, going back to the pioneering British efforts in 1936, show a promise warranting additional attention. The problems of this approach arise mainly from the stickiness, lasting about one second, which develops in the raw coal and complicates the design of equipment for conducting the treatment.

El Paso Natural Gas, in selecting a process for its projected plant in New Mexico, had no choice but the Lurgi gravitating-bed gasifier, a German development of the 1930s, for the initial coal-processing step.

We should perhaps reiterate what is implicit in the foregoing. How very hard it will be to get really serious R&D underway in America for an industry which lacks an R&D tradition and can use only plants of huge capacity. Remember, both the oil industry and Solihull could begin small.

R&D POSTURE OF THE ELECTRICITY INDUSTRY

From its earliest days, the electricity industry has grown at a rate well above that of the general economy. As it grew, it needed bigger machines, and it was compelled to mount a vigorous program of equipment development just to provide such machines. Evolutionary improvements in the machines were achieved at relatively small risk, leading to progressively higher steam pressure and temperatures and higher efficiencies. It was not until the failure in the late 1950s of several plants designed for steam pressures beyond 5000 pounds per square inch and steam temperatures around 1200°F that this evolutionary process came to a halt, and even these failures were relative, for the plants were able to run at lower steam temperatures. After 1960, power engineers agreed that theirs was a mature art, not apt to develop further along the traditional lines. They could take much satisfaction in the high sophistication of their equipment. Many challenging hurdles had been passed in reaching extremes of steam pressure and temperature, in providing enormous steam flows, and in expanding these flows to create power.

Yet all of this depended little upon advances in combustion practice beyond art which was well in hand by 1940, when machines were much smaller and steam conditions far more modest.

Around 1915, power engineers had felt the need for a new combustion technique capable of providing a steam flow larger than their historic grate-combustion devices could readily sustain. Looking around, they found that the cement industry had developed techniques for pulverizing coal and burning the coal powder. Coal pulverizers, coal-conveying equipment, and pulverized-coal firing nozzles were available for purchase from vendors catering to the cement industry. Power engineers found it relatively inexpensive to undertake experiments on pulverized fuel (PF) firing for raising steam. After the successful commissioning of two 20 megawatt turbines in the Lakeside Station in Milwaukee in 1922, PF firing quickly became the choice for nearly all new power-station construction. Notice that the new firing technique was introduced while power stations were still small.

Subsequent advances in coal-firing technology largely involved improvements in equipment for withdrawing ash. The wet-bottom furnace was a development of the 1920s, and the cyclone furnace of the 1930s, each providing means for greater withdrawal of ash matter as a slag and lesser carryover of fly ash to equipment for dust collection.

Research on PF firing did not prove rewarding, for the need to provide sufficient heat transfer surface was usually more important in setting boiler size than the need to provide combustion room. Quite properly, the power engineer concentrated his attention on other problems. As recently as last year, the American Society of Mechanical Engineers, the profession responsible for boiler design, had no standing research committee on combustion (the closest being a committee concerned with fireside corrosion). It should be noted that the ASME is now reorganizing its research committees and will no doubt recognize new responsibilities arising from environmental concerns.

The utility industry and its suppliers were in a poor posture to respond with vigorous R&D to meet the demands which environmentalists placed upon the industry in the late 1960s.

The situation was aggravated by the fact that the industry had grown to depend upon vendors for much of its progress. This was in part because of pressures from regulatory agencies not to undertake research at its customers' expense, but no doubt was also in part because of the industry's essential and demanding preoccupations, providing continuity of service and planning new capacity and raising money therefor. Utility executives were far more conscious than pipeline-gas executives of the role of technology in their affairs, but their experience had been limited mainly to approving fairly modest evolutionary changes on the word of their suppliers. They were not accustomed to making R&D decisions entailing a revolutionary change in practice. The adoption of nuclear power is only an apparent exception in this respect, for it was done not only under urging of the highest scientific authority but also under psychological imperatives: "don't hold up progress," and "if you don't, the government will." A few companies, like Commonwealth Edison of Chicago, made significant early contributions to development of the new power source, but most of the industry was content to "follow the leader."

A mature industry which relies primarily upon its vendors for technological advance is of course in thrall of its vendors' interests. Vendors with a large stake in current procedures have small incentive to examine revolutionary ideas.

It is no wonder that the industry's first response to the environmentalists was to search for devices which could be tacked onto existing boilers. The greatest hope was placed in limestone-injection processes. Unfortunately, the dry injection process has failed, and wet-limestone scrubbing processes appear to be heading for serious trouble.

In February 1970 a panel convened by the National Academy of Engineering concluded that *"contrary to widely held belief, commercially proven technology for control of sulfur oxides from combustion processes does not exist"* (original emphasis). The panel judged that a process must pass the test of operation on the scale of at least 100 megawatts before it could be judged commercially proven. By this test, the panel's conclusion of February 1970 still stood in February 1972!

In spite of lack of proven technology, the electricity industry has committed more than $250 million to installations now under construction for cleaning stack gases. Doubts are arising concerning the operability of several processes, and power engineers are beginning to look elsewhere. A beginning is being made on attempts to rethink the problem of burning coal or residual oil with air pollution as a first consideration, not last.

AN OPPORTUNITY FOR THE ELECTRICITY INDUSTRY

It is often fun to speculate how a technology might have developed under altered circumstances. No doubt much of technology is "inevitable"; it would be hard to beat the wheel. Some things, however, are not inevitable, although an aura of inevitability tends to surround a mature art practiced on a large scale, protecting it from competitive ideas. Depending upon it hang the investments of suppliers and the careers of many men in the employ of both users and suppliers. Such men are poor counsellors on the new.

A major opportunity for the electricity industry has its origins in the 19th century. The industrialist of that era, who had to rely primarily upon coal for his energy, sometimes needed to apply clean heat at high temperature. He could not use the dirty products of combustion of coal for heat treatment of metals, for example, or for making ceramics and fine glassware. For such needs, he resorted to a two-step combustion. He furnished coal to a deep bed blown with air and steam at the bottom, to obtain a fuel gas comprising mainly carbon monoxide, hydrogen, and nitrogen. This he would cool and scrub with water to remove dust. If desired, he could remove sulfur compounds by passing the gas through a bed of iron oxide. He could then burn the clean fuel gas to provide the desired clean heat. The gas had a heating value on the order of 150 BTU per cublic foot, about one-sixth that of natural gas.

Shortly after 1880 (Ayrton, 1881?), he began to use the gas, now dubbed "power gas," in Otto gas engines to make electricity. Well into the 20th century, power engineers could argue convincingly that gas engines were better than steam engines (Robson, 1908). So long as installations were on the scale of only a few hundred or even a few thousand kilowatts, the reciprocating gas engine with its power gas producer was both cheaper and more efficient than a reciprocating steam engine and its boiler.

In light of today's concerns, it is startling to learn that at least a few French gas engines made electricity without emitting either ash or sulfur dioxide, for iron oxide was provided to rid the power gas of sulfur. The general opinion, however, was that this complication did not yield benefits justifying its cost (Robson, 1908; Latta, 1910).

When installations grew beyond a few thousand kilowatts, bringing Parson's new steam turbine into the competition, the advantage both in first cost and in operating efficiency passed to the steam system. Soon, as we have seen, need was felt for a new combustion technique, and PF combustion was developed.

The story of electricity from coal might have been very different if installations beyond several thousand kilowatts had come along 30 years later, after the gas turbine and the art of fluidizing solids were on the scene.

The gas turbine is an inherently cheaper device, even in relatively small sizes, than a steam turbine with its boiler. Alone, the gas turbine is not as efficient, in today's designs, and has not been applied for providing baseload power. Most gas turbines used by the electricity industry are in peakload service and burn expensive clean fuels. In such service, there is little incentive to put to use the heat in the hot gases discharged by the turbine. In baseload service, however, a waste heat boiler may be provided to raise steam from the turbine's exhaust heat. The resulting combination of gas turbine and steam system has advantages of both lower capital cost and higher efficiency than a steam system acting alone. Capital cost is lower for three reasons: part of the power is made by an inherently cheaper machine, the gas turbine; a boiler recovering heat from the clean gases exhausting from the gas turbine can be considerably cheaper than a boiler for a dirty fuel; and the steam turbine working in tandem with a gas turbine need not be supplied with steam at such high pressure for optimum results.

The electricity industry has long been intrigued with the potential advantages of combined gas- and steam-turbine systems. During the 1950s, serious consideration was given to a scheme for carbonizing coal to provide a fuel vapor to a gas turbine. Since this would operate at an air flow several hundred percent beyond the stoichiometric quantity to burn fuel vapor, it would exhaust gases containing oxygen to a boiler fired with the char that remained from the carbonization. It was decided that the advantages were marginal, and the proposal failed of adoption. It is a pity that this was so, for experience with the scheme on a commercial scale would have been valuable to utility executives now facing revolutionary decisions regarding combustion technology. It should be mentioned that large-scale tests of the carbonization step did provide valuable experience in handling coal in fluidized beds.

The decision not to acquire commercial experience using a scheme which represented a major shift in practice must be viewed, however, as characteristic of the industry in recent years. It is a great pity that some user of steam in the U.S. did not have the curiosity to install here an example of the Ignifluid boiler shortly after its development in France in 1955 (Squires, 1970). Its inventor, Albert Godel, had the wit to fluidize coal in a bed resting upon a travelling grate. He found that he could gasify coal with air and a little steam to make power gas at a rate per unit of grate area approaching ten times beyond the rate at which historic grate-combustion devices could consume coal. He discovered that ash-matter of substantially all coals is self-adhering at a temperature in the vicinity of 2000°F. As a coal particle is consumed in the Ignifluid boiler, ash is released. Ash sticks to ash and not to coal, and ash agglomerates form. They sink to the grate, which carries them to an ash pit. For most coals, substantially complete carbon burnup is achieved simply by returning particles which are blown out of the bed. In the Ignifluid, the power gas produced in the fluidized bed is burned in a secondary combustion above the bed.

It is significant that the Ignifluid boiler was developed in a country interested in the technology of small boilers because it is keen to sell such equipment on the world market. This cannot be said of the American boiler industry. More than 35 Ignifluids have been sold all over the world, but none here.

Suppose the gas turbine and the Ignifluid had been available when electricity installations passed beyond sizes attainable by the gas engine. Would PF combustion have then seemed attractive? Remember, the gas engine was clean, and PF combustion generates fly ash. The question is not worth arguing, but gas turbine machinery soon to become available will dispel the aura of inevitability that tends to surround PF combustion and colored early thinking on sulfur dioxide control.

Aircraft gas turbines of latest design operate at 2200°F during take-off and at temperatures not much less while cruising. Land-based machines can now be specified for steady operation at 1800°F, or for operation at 1900°F if they are to be used intermittently to meet peaks in demand for power. The gas turbine is coming of age, and the advantages of combined gas- and steam-turbine systems can no longer be ignored.

Thanks to the needs of the aircraft industry, a major R&D activity in this country is directed toward improving gas turbine art. Land-based machines for steady operation at 2400°F will become available before 1980. United Aircraft has told the Environmental Protection Agency that 2800°F can be reached by 1985 and 3100°F by 1990 (Robson et al., 1970). Machines larger than 100 megawatts can be expected by 1980, and United Aircraft projects sizes of 300 megawatts for 1990. These results will come from the efforts of men now engaged in development and design activities which are funded at several hundred millions per year. We do not speak of a theoretical possibility, but of an almost certain commercial result.

Availability of these machines will create an overpowering economic incentive to find ways to provide them with clean power gas made at high pressure from coal or residual oil.

Fortunately, technologies exist today, not ideal because they were developed for other purposes, which may be used to build experimental installations at once to gain experience in firing gas turbines with coal and residual oil. Like the early gas engines, they will provide electricity with absolutely no emissions of dust, and, as we shall see, they can be fitted at moderate cost with equipment to suppress emissions of sulfur dioxide. Most fortunately, they will emit far less nitrogen oxides than conventional stations, probably less by roughly two orders of magnitude (Robson, 1971). Because of their higher efficiency, they will discharge less waste heat to the environment.

Stations based on existing technologies for gasifying coal and residual oil can provide experience as well as convenient sites for development of better gasification procedures. Let us consider the two fuels in turn.

CLEAN POWER GAS FROM COAL

The gravitating-bed gasifier marketed since 1936 by Lurgi Mineroel-technik GmbH, Frankfurt (Main), West Germany, can be modified to convert coal to power gas at 20 atmospheres. More than 50 units have been built to provide town gas or synthesis gas. They resemble the historic power gas producers save in their ability to operate at high pressure and in their use of oxygen rather than air. To make power gas at high pressure, they would use air. The power gas would be far cheaper than town gas or synthesis gas, let alone gas of pipeline quality. Oxygen is obviously far more expensive than air, and the manufacture of pipeline gas requires an expensive processing train with many steps, while the Lurgi can provide crude power gas in one step.

When the Lurgi is operated to make power gas, coal would gravitate downward in a bed against a rising flow of air and steam, introduced through slots in a rotating grate. Directly above the grate, oxygen in the air is consumed in a shallow combustion zone which converts the last carbon in the solid to carbon dioxide. Ash, typically containing a few percent carbon, is discharged below the grate. Hot gases comprising carbon dioxide, steam, and nitrogen rise upward from the combustion zone through the carbon bed giving up heat to sustain the endothermic reactions of steam and carbon dioxide with carbon to yield hydrogen and carbon monoxide. When the rising gases have been cooled to about 1400°F, these reactions effectively cease. Further cooling of gases is simply by exchange of heat with raw coal. This heat exchange drives methane and tars from the coal and typically cools the gases to about 950°F. The gases are then quenched with water to reduce their temperature to about 320°F.

Sulfur compounds may be scrubbed from the crude power gas by any one of several alkaline scrubbing liquors at a cost far below the cost to scrub sulfur dioxide from stack gases of a conventional power station. The reasons for the lower cost are fairly obvious. Chemistries for absorbing the hydrogen sulfide present in power gas are freer of disturbing complications than chemistries for absorbing sulfur dioxide. Hydrogen sulfide can be converted more readily to elemental sulfur, the only product which a power station might successfully market over the long run, or stockpile in absence of markets. Power gas from the Lurgi comprises only about 40% as many moles as stack gas, and because the power gas is at 20 atmospheres, its volume is only 1.7% of stack gas.

Lurgi is now building a pioneering installation for Steinkohlen-Elektrizitaet AG at Luenen, West Germany, in which five Lurgi gasifiers will supply power gas to a gas turbine for 74 megawatts. Associated with the gas turbine will be a steam turbine for 98 megawatts.

The Lurgi gasifier has been used on a wide range of coals, including weakly caking bituminous coals, but may not be suitable for processing some strongly caking eastern U.S. coals. Commonwealth Edison Company of Chicago has engaged Lurgi to perform engineering studies of systems to supply clean power gas made from Illinois coal, and one aspect of the studies will be to determine the suitability of moderately caking Illinois coal for the Lurgi gasifier.

As an immediate solution of the pollution problems of an existing steam power station, Lurgi proposes that clean power gas made at 20 atmospheres be let down in pressure through a turbine generating a relatively small amount of electricity and then used to fire the station's boiler.

It is necessary now to deal with the Lurgi gasifier's faults. Admirably suited for its original purpose, manufacture of town gas or synthesis gas, the Lurgi suffers the serious disadvantage that the products of combustion of Lurgi power gas will contain a large amount of water vapor. This is inherent in its principles of operation. In order that it may discharge a loose, nonagglomerated ash, a large amount of steam must be supplied with the combustion air to keep down the combustion zone temperature. Most of this steam is converted to hydrogen in the endothermic gasification zone above the combustion zone. Although hydrogen was a desirable constituent of town gas or synthesis gas, this fuel species gives rise to an undesirable release of water vapor from the stack of a power station using Lurgi power gas.

Crude Lurgi power gas contains tars and chemically active species which would polymerize to form tars if not cooled quickly. Another undesirable input of water vapor arises from the necessity to quench the crude power gas. About 5% of the sulfur in the crude gas is in the form of organic sulfur compounds difficult to remove.

The Lurgi must be fed with coal from which fine particles smaller than about ⅛ inch have been removed. When coal is mined, coal fines are inevitably produced. If the Lurgi is to utilize run-of-mine coal, a pelletizing or agglomerating step must be provided to deal with these fines.

The Lurgi has a limited coal-processing capacity. The Luenen installation includes five coal-gasification vessels 13 feet in outside diameter to provide power gas for 170 megawatts. Scale-up of the Lurgi to larger capacities may prove difficult and uncertain.

A team at The City College of New York, supported by grants from the Environmental Protection Agency, has been studying an alternative which can serve here to illustrate the possibilities for improvement upon the Lurgi (Graff et al., 1971).

The fluidized bed provides an attractive technique for bringing run-of-mine coal, merely crushed to a size smaller than about ¾ inch, into intimate contact with air and steam. In a fluidized bed, rising gases buoy granular material, setting it into motion. Large-scale movement of the solid conveys heat from exothermic zones (such as a combustion zone) to endothermic zones (such as a zone for reaction of steam and carbon dioxide with carbon), and the temperature of a fluidized bed is uniform.

A single fluidized-bed reaction vessel could easily provide power gas for 1000 megawatts. A fluidized-bed gasifier can deliver gases free of tars and chemically active species which would polymerize to tars. No sudden quench of the gases would be required. Steam supplied directly to the gasifier can be a small fraction of that needed for the Lurgi, because combustion heat generated near the air inlet is carried away by motion of the solid.

The City College team is examining a gasifier which is essentially the ash-agglomerating fluidized bed of the Ignifluid boiler redesigned for operation at high pressure (Squires, 1971). The team is studying a technique for ridding the crude power gas of dust and sulfur by means of a filter using a granular solid derived from naturally-occurring dolomite rock. Half-calcined dolomite, containing calcium carbonate and magnesium oxide, has been found to be extraordinarily reactive toward hydrogen sulfide (Ruth et al., 1971). The solid may be regenerated with conversion of sulfur to elemental form for the market, allowing the solid to be used repeatedly (Squires, 1967). A granular solid disposed in a panel bed filter (Squires and Pfeffer, 1970) can filter dust from gases at efficiencies beyond 99.9%. The potential advantage of The City College gas-cleaning scheme is that it can be applied to hot power gas, obviating the necessity to reduce its temperature to 320°F or below in order to apply a scrubbing technique.

The accompanying Table illustrates both the importance of reducing water vapor in power gas and the advantage of hot cleaning.

The first column of the Table tabulates a representative energy balance for conventional steam power equipment without sulfur recovery. The second column is an energy balance for the state-of-the-art Lurgi gasifier, followed by state-of-the-art gas cleaning at low temperature, supplying power gas to a gas turbine working at 2800°F, a level which United Aircraft has stated can be attainable by 1985. The gas turbine is followed by a conventional steam system working at 2400 pounds per square inch and 1000°F with one reheat to 1000°F and exhausting at 101.4°F. Comparison of the first and second columns of the Table shows the incentive, an improvement from 39.5% to 45.5% overall efficiency in power generation, for R&D on power-generating machinery to couple with state-of-the-art gasification equipment.

The third column is an energy balance for the ash-agglomerating fluidized-bed gasifier, followed by state-of-the-art gas cleaning at low temperature, supplying gas to power generating equipment identical to that postulated in the second column. Note the sharp reduction in loss of latent heat to the atmosphere. Comparison of the second and third columns measures the incentive for work on a gasifier better than the Lurgi, an improvement from 45.5% to 49.7% efficiency.

The fourth column is an energy balance for the ash-agglomerating fluidized-bed gasifier, followed by The City College scheme for cleaning gas at high temperature, followed by the identical power-generating equipment. Comparison of the third and fourth columns roughly states the incentive for work on techniques for hot cleaning, an improvement from 49.7% to 51.2% efficiency, although it appears that the hot-cleaning procedure will also enjoy lower capital cost.

Table 1 Illustrative Energy Balances

	Conventional Steam Power Equipment, Without Recovery of Sulfur	Advanced Power-Generating Machinery Combining Gas and Steam Turbines, with Recovery of Sulfur		
		Lurgi Gasifier, Gas Cleaning at Low Temperature	City College Ash-Agglomerating Fluidized-Bed Gasifier	
			Gas Cleaning at Low Temperature	Gas Cleaning at High Temperature
Electricity sent out	39.5	45.0	49.1	50.5
Heating value of sulfur	—	1.0	1.1	1.3
Loss of sensible heat to atmosphere in stack gases	5.0	4.6	4.5	4.7
Loss of latent heat to atmosphere (water vapor in stack gases)	3.8	14.1	5.6	4.5
Rejection of heat to environment at steam condenser and elsewhere	47.7	28.4	35.7	35.0
Estimate of loss of unburned fuel, leakage of heat from equipment	2.0	4.9	2.0	2.0
Estimate of mechanical losses and power taken by auxiliary equipment	2.0	2.0	2.0	2.0
	100.0	100.0	100.0	100.0
Efficiency of conversion of fuel energy to electricity, allowing credit for heating value of sulfur	39.5	45.5	49.7	51.2

Basis: Coal input = 100.

The hot-cleaning approach may fail if emissions of volatile materials, such as mercury, prove critical and if means for dealing with these substances at high temperatures cannot be found. It is too early to dismiss hot cleaning for this reason.

Another technique for gasifying coal which must be considered is the slagging gasifier studied on a small scale during the 1950s by the Bureau of Mines and the Institute of Gas Technology, and said to have been tested on a large scale by Texaco in about 1957, although nothing was published. To make power gas by this approach, finely powdered coal would be reacted in a refractory-lined chamber with air and steam at high pressure. Discussions with the Texaco workers led to the impression that their philosophy was to use a number of small units if a large capacity was desired, and it may be that the slagging gasifier does not lend itself as readily as a fluidized bed to design for a large scale. Other questions to be answered are refractory life and efficiency of utilization of carbon. Data published on small-scale work, as well as experience of commercial slagging gasifiers working at atmospheric pressure, suggest that carbon utilizations beyond 99%, normal for PF combustion as well as the Ignifluid, may be difficult to attain. Efficiency of utilization of steam is almost certainly less than that afforded by an ash-agglomerating fluidized bed, and results for the slagging gasifier at best will probably fall about one efficiency unit below results shown in the Table for the ash-agglomerating bed. The City College hot-cleaning technique could be applied. On the whole, the slagging gasifier appears to present tougher problems than the ash-agglomerating bed, and may be viewed as being not much farther along in its development if the latter be regarded as an evolution from the Ignifluid.

Unfortunately, work conducted to date by organizations under contract to the Office of Coal Research and by the Bureau of Mines with the object of producing SNG (i.e., substantially pure methane) will not contribute much toward development of power gas processes for the electricity industry. The former is a more difficult task, for which it is desirable to adopt a processing scheme which maximizes the quantity of methane made directly from coal. Schemes under study all provide several steps for contacting coal with gases, the flows of solid and gases being counter-current. Such schemes are not useful for making power gas. The exact heating value of power gas does not matter very much, and the objectives should be simplicity, the lowest possible cost, and highest possible steam and carbon efficiencies.

CLEAN POWER GAS FROM RESIDUAL OIL

Both Texaco and Shell have marketed a process in which oil is reacted with oxygen and steam at high pressure to furnish synthesis gas (hydrogen and carbon monoxide) for conversion to ammonia. An experimental installation in which the oxygen is replaced by air could provide early experience in firing a gas turbine with power gas made from oil. The process suffers from a high yield of soot. Experiments could advantageously be conducted on hot-cleaning techniques, like the one under study at The City College, for capturing soot, so that it may be recycled to the reaction chamber. Soot from an oil of Caribbean origin will contain vanadium and might be marketable to earn more than $1 per pound of vanadium pentoxide context. The return to an operation using a bottom cut of a vacuum distillation of Venezuelan residual oil could run beyond 50¢ per barrel, far greater than the roughly 10¢ return from sale of byproduct sulfur.

Ube Industries, Ltd., of Japan has worked on a process in which oil is severely cracked at essentially atmospheric pressure in fluidized beds, heat being supplied with oxygen, to obtain a gas of about 800 to 900 BTUs per cubic foot which can be desulfurized by a scrubbing technique. Ube has worked on the scale of both one and five tons per day. The Japanese government is said to have decided to grant a subsidy toward erection of a semi-commercial plant to supply 10 million cubic feet per day. It is doubtful that a process working at atmospheric pressure and depending upon oxygen can be best in the long run, but it is sad to see Japan get this head start. A similar installation here under a license could provide valuable teaching.

An argument can be made that work on residual oil should take on higher priority at the outset than work on coal. We will return to this point shortly.

A REMARK ON MAGNETOHYDRODYNAMICS

We have seen that gas turbines will arrive shortly permitting design of power systems for efficiencies beyond 50%. The gas turbines would work in cooperation with steam cycles which scavenge heat from the hot gases emerging from the turbines.

The need to provide two types of power-generating machinery follows from fundamental principles of thermodynamics and the character of the machines. The Carnot principle of thermodynamics teaches the power engineer that he should, ideally, introduce heat into his power-generating medium—combustion products in the gas turbine and steam in the steam turbine—at as high a temperature as possible; also, that he should withdraw heat from the medium at as low a temperature as possible after its expansion in a power-generating device. A gas turbine is ideally suited to meet the first requirement, poorly suited for the second. The steam turbine is ideal for the second and poor for the first.

A combined system enjoys the advantages of both: the gas turbine first extracts work from high-level heat, passing on to the steam system the leftover heat, now at an intermediate temperature level. The steam turbine extracts work from heat at lower temperature level, and rejects heat to the environment at the lowest possible level, just that little bit above the temperature of the environment which is necessary to cause heat to flow across the metal walls of the tubing of the steam condenser.

An alternative opportunity for power systems with efficiencies beyond 50% has received much popular attention. This is the magneto-hydrodynamic (MHD) technique for power generation. Like the gas turbine, an MHD generator must be followed by a steam cycle scavenging heat that the MHD generator cannot use. Like the gas turbine, the MHD generator is a device for extracting work from heat at high temperature levels.

From the commercial standpoint of electricity industry, the gas turbine has the enormous present advantage of being much farther along in its development. Small test machines for temperatures as high as 3100°F have been in operation for at least the past decade. Larger test machines for temperatures in the range from 2400° to 3100°F have been studied. Commercial designs for machines in the vicinity of 2400°F are in the works. These machines are the fortunate spin-offs from efforts to make better aircraft. They will arrive on the commercial scene with little or no help from the power industry or the federal establishment, beyond help already given toward better military turbines.

It is hard to escape the impression that the gas turbine has too much of a headstart for the MHD device to overcome. The latter's development is fraught with difficulties, although it would be only fair to add that so also might have seemed the development of 2000°F gas turbines in 1950.

From an environmental standpoint, the gas turbine may be superior. United Aircraft (Robson, 1971) believes that exhaust from a gas turbine burning power gas will contain very little nitrogen oxides. The low heating value of the gas leads to a lower flame temperature in the central combustion zone of the gas turbine flame, where the fuel-to-air ration is substantially stoichiometric. The nitrogen oxides level is lower at the lower flame temperature, whatever the final temperature of the gases entering the turbine. It should be remarked, however, that the Bureau of Mines is working on a combustion system for MHD which Bienstock, Demski, and Demeter (1971) believe can reduce nitrogen oxides emissions to acceptable levels, as well as capture sulfur.

It would be fair to say that the MHD generator has attracted more enthusiasts than the gas turbine. Their energetic advocacy has established MHD in the minds of many as the main chance.

It should be said that gas turbine makers have been busy in recent years meeting the sharp increase in power industry demand after the blackout of 1966 and the nuclear delays. They have had little time or energy for promoting their long-term interests. Conversations with some of these men have led to the impression that their managements have been slow to see the opportunities. Companies manufacturing large stationary gas turbines have far larger investments in the traditional steam technology. The companies have been deeply engaged, moreover, by the problems of launching nuclear technology. They appear to have been slow to assign personnel to the study of long-range implications of advances in gas-turbine art, let alone personnel to the tasks of advocacy.

A REMARK ON IMPLICATIONS OF THE NEW POWER TECHNOLOGY

Development of new power systems incorporating either gas turbines or MHD generators will have important implications for all of the fossil fuel industries.

A system with a gas turbine must inevitably include equipment for processing coal or residual oil to furnish a clean fuel to the turbine. Although a system with an MHD generator must have cleaning equipment on the exhaust side of the generator in any case, to recover the seed substance which has rendered gases electrically conductive, it would appear that equipment for ridding the fuel of ash and sulfur ahead of the generator will be cheaper to provide than equipment on the exhaust side. This is simply because the processing of the dirty fuel will be easier if it is done at an elevated pressure.

Presence of a step in which the dirty fuel is treated at high pressure opens up the opportunity for process modifications which will "skim off cream" in form of fuel products of higher value. In the long run it will be a shame to use raw coal or even residual oil for generating electricity. Chemically-bound hydrogen in these fuels is too valuable simply to burn to steam and send up a stack as water vapor. In the end, power generation must be based on a coke residue from an operation in which valuable gaseous and liquid fuels have been extracted.

As energy costs rise, the incentive will grow to hunt for savings which can result from combining several objectives. An installation to convert coal or residual oil to SNG, clean liquid fuel, and clean electricity can be both cheaper and more efficient than three installations to make each product separately. The installation might advantageously provide low-sulfur coke for metallurgy and electrochemistry. It might supply feedstocks for petrochemicals. It might distribute an "industrial gas" to industry and older power stations, working at low load factors, in districts such as Pittsburgh, Chicago, Cleveland, Philadelphia-Camden, North Jersey, and Houston. The heating value of industrial gas would lie intermediate between that of power gas and SNG; so would the cost.

At The City College, we call such an installation a "Fuelplex," and we are examining examples of both the "Oilplex" and the "Coalplex" as a guide to picking the best fuel-conversion processes for development. Developments will be more interesting the greater their potential of evolving into operations useful in a Fuelplex.

The Fuelplex provides the chemical engineer with a new set of rules:

1) Air at high pressures will be available "free" from the compressor supplying air to the combustion for the gas turbine or MHD generator.
2) Processes for making SNG or liquid fuel may freely reject a gas at high pressure as a byproduct, even a fuel gas so lean as normally to be considered incombustible, since the rejected gas can be used by the gas turbine or MHD generator.
3) The processes may reject heat at high temperature levels without the serious loss of efficiency which such a procedure would normally entail, since the steam system which scavenges heat from gas turbine or MHD generator exhaust can also scavenge from a chemical operation. This opens up a range of new high-temperature chemistries for use in the Fuelplex.

These new rules provide the chemical engineer with an invitation to invention, and there can be little doubt that the power station of the future will become a scene for exciting new chemical operations leading in general to greater efficiency of utilization of fuel values. In this context, removing dust and sulfur from power gas supplied to a gas turbine or MHD generator will seem a mere incidental.

The gas turbine has an advantage over the MHD device for the Fuelplex, since operation of the latter is inherently limited to a pressure on the order of six atmospheres. The best pressure level for a gas turbine is higher at higher temperatures, and turbines for operation at around 20 atmospheres can be expected soon. The higher pressure is a better level for fuel processing. Higher pressures will generally favor higher yields of SNG and liquid fuels, and gas turbine art is confronted with no inherent limitations preventing choice of the best pressure which balances its own economics against the economics of fuel products for distribution.

Immediate short-range development opportunities exist for both coal and residual oil, especially the latter, which can launch efforts leading to the ultimate Fuelplex. Let us again consider the two fuels in turn.

OPPORTUNITIES FOR RESIDUAL OIL

Opportunities for residual oil are best viewed in light of world markets. Our declining balance of payments should teach us to pay more attention to world needs in selecting R&D goals, for we are rapidly losing ground as an exporter of technology. Indeed, we are importing technology, as we have seen earlier in the discussion of processes for SNG from light naptha and gas oil.

Recent years have seen a marked shift toward residual oil as the preferred fuel in new facilities for providing electricity or industrial heat in developed economies. Coal has almost disappeared as an item of world commerce except for shipments of coal for metallurgy. Developing countries lacking supplies of native coal depend almost exclusively upon oil for energy.

The reasons for this are simple. Residual oil is produced as an incidental to production of liquid fuels for transport and other uses, it is cheap to carry on the seas, and is convenient to use.

Most residual oil shipped on the world market is the bottom cut from a distillation at atmospheric pressure. This cut is often a substantial fraction of the oil leaving the ground. For example, the atmospheric bottoms from Venezuelan crude amounts to about 57% of the total.

In the U.S., on the other hand, refiners convert as much native crudes as possible to products of higher value. The atmospheric bottoms is distilled under vacuum to obtain a heavy gas oil suitable for cracking to lighter products. The vacuum bottoms is often subjected to further processing, either by solvent extraction or by a deep vacuum distillation carried to 1400°F. Some of the vacuum bottoms is often destroyed by a coking operation yielding lighter products and a petroleum coke, which enters either the fuel market or technology for production of electrodes and other carbons of higher value. In recent years, the residual oil finally remaining has amounted on the average to about 7% of the oil charged to American refineries.

Residual oil is nevertheless a major source of energy, especially for the East Coast. About one-half of the oil burned in the U.S. is consumed in five states: New York, New Jersey, Pennsylvania, Massachusetts, and Florida. Imports furnish about 60% of the residual fuel oil burned domestically, and imports have nearly doubled their share of the market in the past decade, as domestic refiners reduced their yields of heavy oil. In 1966, Venezuelan sources supplied about 90% of the heavy oil burned on the East Coast.

About one-quarter of the oil is used to generate electricity, one-quarter for heating, and one-quarter for industry. Most users of oil for heating and industry purchase oil in such small amounts that the only practicable way for these users to meet regulations limiting emissions of sulfur dioxide will be to obtain oil of low sulfur level from fuel suppliers. A large power company, on the other hand, buys oil in tremendous quantities, and this fact suggests that the economic answer to this company's sulfur dioxide emission problem may lie elsewhere (Alpert et al., 1971). Recent economic developments strongly reinforce this suggestion.

Two years ago residual fuel oil at 2% sulfur sold for $2 per barrel; fuel oil at 1% sulfur sold, in relatively minor amounts, for about $2.50; fuel oil at 0.3% sulfur was, for all practical purposes, unobtainable. Today, fuel oil at 2% sulfur costs $3.50 per barrel, the same as domestic crude. Fuel oil is available containing 0.3% sulfur at a price of $5. Apparently no petroleum economist anticipated these price responses to the rapid growth in demand for fuel oil and governmental pressure for reduction in emissions of sulfur dioxide.

The cost to deliver Middle East crude to the U.S. would be only $2.80 per barrel if a suitable port were available here for supertankers.

These prices have made it attractive to import oil to the U.S. East Coast with object of producing from it just two main products, SNG and fuel oil of low sulfur content. Crown Central Petroleum Corporation wishes to build a plant for these two products in Baltimore. SNG would be made by the processes mentioned earlier which use distillate petroleum feedstocks. The project requires approval by the Oil Import Administration, which is studying similar proposals from several other companies. Although details are not available, it would appear that the plan is to sell some oil at 1% sulfur as well as oil at 0.3% sulfur. It will be necessary to discuss briefly the techology for desulfurizing residual oil to shed light on a problem which can arise if all of the oil must be shipped at 0.3% sulfur.

Oil companies have already made sizeable investments in equipment to desulfurize residual oil. Most installations have used an "indirect" technique. In this approach, the atmospheric bottoms is distilled under vacuum to remove a heavy gas oil fraction. This is contacted by hydrogen at high pressure over a catalyst, to convert a large portion of the sulfur in the fraction to hydrogen sulfide and to leave an oil containing about 0.2% sulfur. If a product of about 1% sulfur is acceptable, the desulfurized fraction can be blended back with the residue from the vacuum distillation. Often the blend is further cut with low-sulfur oil from another source, such as low-sulfur North African crude.

The indirect technique is well suited to reduce the sulfur level of Caribbean residue to about 1%. It is poorly suited for achieving lower sulfur levels, which would require treatment of the bottoms from the vacuum distillation. For Caribbean oils, this is difficult, because they contain vanadium and nickel in amounts which would poison catalysts used in the hydrodesulfurization process.

We have already mentioned the "race" by nine teams to develop "direct" processes for desulfurizing the entire atmospheric bottoms in one operation. Unfortunately for the developers of these processes, it would now appear that at least 60% of the residual fuel oil market in the U.S. in 1980 will require oil levels of about 0.3%. The direct processes do not now appear capable of providing such a product, or indeed anything below about 1% sulfur. The present view is that material at the 0.3% sulfur level can be made only from a distillate stock. This means that providing oil at 0.3% sulfur will leave large quantities of pitch undealt with. It will not even be possible to burn these "dregs" within stateside refineries, in view of pollution regulations, and its accumulation will create a disposal problem.

The refiner's problem will become the power industry's opportunity. It should be able to purchase the refiner's dregs at a large discount, and it will therefore have a strong incentive to learn how to use this material to fuel gas turbines.

Crown Central's proposed Baltimore installation could be converted to an Oilplex by addition of Texaco or Shell partial oxidation to convert pitch to power gas for gas turbines. This addition could greatly ease some problems faced by the installation's designers. The pitch typically amounts to 20% of the crude, and would contain 5–6% sulfur if derived from Kuwait crude oil, or 4% if derived from Venezuelan crude. The quantity of pitch is greater than that which might be used to make hydrogen for converting lighter distillate materials to SNG and for desulfurizing heavier distillates. Attempts are in progress to find ways to reduce the quantity of pitch by deasphalting procedures; its commitment to power generation is an attractive alternative.

An Oilplex could choose from a wider range of crudes, including materials of low present economic value, containing more than 50% pitch of high sulfur content.

The City College team is examining several candidate developments for further evolution of an Oilplex. Two examples will illustrate the possibilities.

The shifting of carbon monoxide to hydrogen by action of steam and lime was a popular development objective before about 1930. Steam reacts with carbon monoxide to furnish hydrogen and carbon dioxide, which is promptly removed from the gas mixture by reaction with lime to form calcium carbonate. Process men lost interest in this approach when they realized that they could not afford to waste the relatively large amount of heat that evolved from the reaction of carbon dioxide with lime. In an Oilplex, this heat could be used to raise steam for production of electricity. Alternatively, it could be used to sustain the cracking of pitch, preferably in presence of a substantial partial pressure of hydrogen. In the latter use of the heat, the lime can serve the function of absorbing sulfur as well as carbon dioxide.

We are examining the role in an Oilplex of a coke-agglomerating fluidized bed for converting pitch to products of higher value. If the bed is fluidized with hydrogen at about 1400°F and 50 atmospheres, it could provide a gas readily upgraded to SNG, a light aromatic liquid fuel, and coke beads low in sulfur and about $\frac{1}{4}$ inch in size. If the bed is fluidized with products of the partial combustion of oil with air, at about 1400°F and 20 atmospheres, it could provide a fuel gas having a heating value about one-half that of methane along with the same other two products.

The latter embodiment could be exploited to meet a number of the energy needs of a developing country. As well as electricity, these include needs for motor fuel, for a fuel gas supplying urban areas, and for a smokeless solid fuel for distribution to rural districts. A low-volatile solid fuel would also have value for supporting small-scale metallurgy. The coke-agglomerating fluidized bed appears well suited to be the sole major oil-processing unit of a "mini-refinery" to serve a developing country. If a world market could be developed for the "dregs" of the refineries of developed economies, the developing country possessing such a mini-refinery would enjoy energy costs well below the world average.

Keith (1971) believes that some 100 billion barrels of oil of high gravity, known to be in place in the ground but not included in our reserves of "producible oil," might be recovered by injecting carbon dioxide at high pressure. Carbon dioxide is soluble in such oil and reduces its viscosity typically by a factor of ten. The viscosity reduction combined with the imposed pressure gradient can produce a dramatic increase in the flow of oil. Keith has increased the yield from a watered-out field in Arkansas from 2200 to 16,000 barrels per month. SACROC plans to spend $175,000,000 to inject carbon dioxide into the Snyder County field in West Texas, and expects in ten years to recover 156,000,000 barrels of incremental oil which could not otherwise be produced. Over the life of the field, incremental recoveries should be much greater. A number of other smaller projects are in progress.

This development deserves the closest governmental attention. It will be easier and quicker to set up plants to make SNG from heavy oil than from coal, as well as easier to furnish clean power gas to gas turbines. An Oilplex making SNG could be designed to supply carbon dioxide gas for injection into the field that provides the Oilplex with its feedstock.

It need hardly be pointed out that money invested in this opportunity—for experimentation with injection of carbon dioxide, development of processes useful in the Oilplex, and erection of a prototype installation—could, if all goes well, improve our balance of payments by reducing our dependence upon imports of oil. Early assessment of the prospects is in order.

In any case, development efforts to convert petroleum pitch to SNG, electricity, and other products should prove more rewarding in the near-term than work on coal. Pitch is an easier feedstock, and commercial results can be expected sooner. The results will have worldwide importance and can lead to income from export of technology. As we shall see, there are few customers for advanced coal technology. Better methods for dealing with refinery dregs can help our balance of payments in two ways: we will be able to make more efficient use of each barrel we must purchase, and we can buy cheaper crudes of higher pitch content. Experience working on processes for pitch can help the effort on coal and might, indeed, speed it along.

These matters are too important to be left to the petroleum industry. As we noted earlier, this industry has shied away from the problems of the regulated fossil fuel industries, gas and electricity. The future of fossil fuel technology belongs to ideas which combine the interests of all three industries, so that advantage may be taken of synergisms which will inevitably result from pooling their objectives. It is unlikely that many petroleum company executives will give much careful thought very soon to hunt for these synergisms, let alone approve the expensive and risky development of the new materials-processing technologies which will be needed to exploit them.

The bright spots in the picture relate to the electricity industry. The ongoing R&D on gas turbines will create an evident need for clean fuel. Executives of large power companies understand that a process trial must be big. By spending more than one-quarter billion dollars on trials of stack-gas cleaning procedures, they have shown their willingness to commit large sums toward a clean environment. It is reasonable to hope that they may take the lead in exploiting the synergisms which can arise when their own needs are pooled with those of men who supply pipeline gas and liquid fuels.

OPPORTUNITIES FOR COAL

From the global standpoint, the use of coal has remained relatively static through the 1960s. Large increases in the use of coal are unlikely in the near future. Coal's importance is fast declining in many areas, notably in northwest Europe. The use of coal may continue to grow at a low rate in the U.S., Russia, Poland, Japan, Australia, and South Africa, but even this is not certain. Coal is important in China and North Korea.

Developing countries have tended to make little use of native coal supplies, and the extent of such supplies is often poorly known.

Keen interest in technologies for suppressing pollutants from the combustion of coal will be found, therefore, only in a relatively few nations having developed economies. In the world market, technologies for oil are much more important.

This is of course not to say that technologies for coal are unimportant here. On the contrary, means for providing clean power from coal are urgently needed. A major shift from coal to oil for electricity production, forced by the tardiness of such means, would be a major disaster, both for our balance of payments and for the economic welfare of our coal-producing districts.

We have already discussed procedures for processing coal to provide clean power gas to a gas turbine. Hottel and Howard (1971) have recently described the processes under study by the Office of Coal Research and the Bureau of Mines for manufacture of SNG from coal.

An experiment underway at Leatherhead, England, should be mentioned. It is supported by the Environmental Protection Agency. Coal is burned completely, in one step, in a fluidized bed operating at 1500°F or above and at elevated pressure. The concept is that offgases from the bed shall enter a gas turbine. Boiler tubes pass through the bed to remove about 70% of the coal's heating value to steam. Experiments in which dolomite was injected to capture sulfur gave strikingly better results than similar experiments conducted at atmospheric pressure.

The Leatherhead experiment must be viewed as opening up a serious possibility for equipment valuable for the middle term. Over the long term, schemes based upon gasification have greater appeal, since only they can permit full utilization of the higher efficiency and lower cost of the expected combinations of gas and steam turbines. In a flowsheet based upon the Leatherhead experiment, the gas turbine at best can provide only about 20% of the total power output, whereas in advanced combinations the gas turbine can support well beyond 50% of the load.

Studies are in progress for processes in which low-sulfur fuels would be made by contacting coal with hydrogen at extreme pressures and in the neighborhood of 850°F.

In one version, the "H-Coal" process of Hydrocarbon Research, Inc., the coal would be treated catalytically in a process closely resembling processes for hydrodesulfurization of residual oil. The product would be a heavy oil much resembling residual oil after the coal's ash matter has been filtered out. It may be remarked that coal in some respects is an easier feedstock than a heavy pitch for such a process. The number-average molecular weight of the individual chemical species present in a low-volatile bituminous coal is typically about 3000; the bulk of such a coal comprises species having molecular weights falling in the range between about 2000 and 12,000; the highest molecular weight might typically be about 50,000 (Sternberg et al., 1970). A pitch generally contains species having molecular weights running into the millions. Such species encounter difficulty entering pores in the interior of a catalyst and can form troublesome smears on the surface of a catalyst particle.

In another version, under development by Pittsburgh and Midway Coal Mining Co. (a subsidiary of Gulf Oil Co.), coal would be first suspended in a suitable solvent, heated in presence of hydrogen, cooled and filtered from its ash matter, and finally removed from the solvent to furnish a product in form of a solid pitch, termed "solvent-refined coal."

It does not appear that such processes can compete with coal treatment which is ancillary to power generation itself. Work on the processes may well be justified with aim of providing fuels low in sulfur to small customers unable to provide means for sulfur control, although it seems quite unlikely that fuels low in sulfur can be furnished such customers from coal as cheaply as from residual oil.

At The City College, we are investigating flowsheets for a Coalplex which we believe will be able to ship SNG, a light aromatic liquid, and electricity at a total cost below the summed costs if each product were made separately.

Schroeder (1962) has pointed the way to a strong candidate for the first coal-processing step in a Coalplex. He reported extraordinary yields of methane and a light aromatic liquid from experiments in which coal was rapidly heated to 1500°F in flowing hydrogen at high pressure. The crucial factor was a short residence time for the vapor products. If these products were allowed to remain at 1500°F for much time, the liquid species reverted to a heavy tar and coke, and so liquid yield declined drastically. On the other hand, if the residence time of vapor products can be held below about five seconds, Schroeder's data indicate a good possibility for yields as high as 40% methane and 30% benzene-toluene-xylene, expressed as weight percent of the starting coal.

To understand Schroeder's remarkable data, we begin by remembering the relatively low molecular weight of chemical species present in coal, cited earlier. If a bituminous coal is heated slowly, it melts, generally at about 750°F. The melting of the coal excites chemical activity, and the mass quickly polymerizes to chemical species of far higher molecular weight, generally in the millions, and it solidifies. Further heating carbonizes the solid mass, but does not yield gas or liquid in the amounts reported by Schroeder. On the other hand, if coal is heated quickly to temperatures on the order of 1500°F, the chemical species in the coal crack to form light materials which pass into the gaseous state. In the first instant at high temperature, as much as three-quarters of the original coal may appear in the vapor state. If the vapor is promptly reduced in temperature, as proposed by Schroeder, good yields of light products can be obtained. Not so if the vapor is allowed to remain at high temperature too long. This is because vaporization of the coal also excites chemical reactivity. Active vapor species quickly polymerize to form molecules so large that they return to the liquid state, forming a pitch. This cracks to form coke and light gases, and the overall effect is a drastic reduction in yield of light products from that obtainable if vapor residence time had been kept short.

A problem in carrying out Schroeder's chemistry is the stickiness, cited earlier, which appears when coal is heated quickly in hydrogen to high temperature. This rules out use of an ordinary fluidized bed of finely-divided particles working at velocities on the order of a few feet per second. Such beds promptly defluidize if raw coal is added. Many proposed processes for treating coal have postulated such beds, however, but first expose the raw coal to a gas containing a little oxygen, to oxidize the coal lightly and thereby to destroy its propensity to become sticky. It should be noted that this pre-oxidation has the disadvantage of reducing the yield of valuable light products from the subsequent coal treatment.

The City College team is examining a coke-agglomerating fluidized bed as a device to exploit raw coal's stickiness. This bed is well suited for conducting Schroeder's chemistry. We would provide a fluidized bed of large carbon particles—from any convenient source at the beginning of the operation—roughly $\frac{1}{8}$ to $\frac{1}{4}$ inch in diameter. The velocity of the fluidizing gas would be on the order of 10 feet per second. We have found that the particles of such a bed can carry a relatively large amount of sticky matter upon their surfaces, without harm to the operability of the bed. In experiments at room temperature, we find that a bed of $\frac{1}{8}$ inch beads fluidized at 10 feet per second can "absorb" an astonishingly large amount of an extremely sticky and viscous material (prepared by dissolving Separan in glycerine) before the bed's fluidization can no longer be maintained. In a comparable bed for Schroeder's chemistry, a freshly introduced particle of finely ground coal would become sticky and adhere to one of the far larger beads of coke, a product of the operation. The fresh material would remain sticky about one second. At reasonable assumptions concerning the throughput of coal, we find that the content of sticky matter in our postulated coke-agglomerating bed would remain far below levels which we achieved in our experiments at room temperature. Perhaps more convincing than our experiments is the fact that Dorr-Oliver has successfully commercialized a number of agglomerating fluidized-bed processes, for calcining calcium carbonate sludges from water treatment and for roasting certain ores, for example. Germany has an agglomerating fluidized-bed process for treating ferrous chloride pickle liquor. We have already cited the ash-agglomerating fluidized bed of the Ignifluid boiler.

An early version of a Coalplex employing a coke-agglomerating fluidized bed might advantageously provide clean baseload electricity and ship a coke low in sulfur to other power-generating sites (Squires, 1970; Squires et al., 1971). Since the hydrogen used to promote desulfurization of the coke would arise autogeneously within the process, no separate facility to supply hydrogen would be necessary. Establishing the feasibility of this approach would seem to be a matter of some urgency, for it could provide a fuel of low sulfur content at a cost below that for synthetic residual oil or solvent-refined coal made by processes working at far higher pressures and needing a facility for hydrogen.

We are also examining the "fast fluidized bed," a recent German development (Reh, 1971), as a technique for conducting Schroeder's chemistry on western coals, which have less tendency to become sticky. In the fast bed, a fine powder is fluidized at a velocity much higher than would have previously been considered attractive for the dense-phase fluidization of such a solid. The trick is to provide a large cyclone and a system for recirculating powder to the bottom of the bed. Powder flows at a large throughput upward through the region occupied by the bed, which lacks a sharply defined upper surface. For a solid resembling the petroleum industry's fluid cracking catalyst, working at a velocity of about 10 feet per second, the fast fluidized bed can maintain a mean concentration of solids on the order of 5 to 10 pounds per cubic foot. Although there is a substantial gradient in solids concentration as a function of height in the bed, the temperature uniformity is comparable with that of conventional fluidized beds.

The fast fluidized bed may also be useful for calcining the calcium carbonate which forms in a process in which lime "accepts" carbon dioxide, such as the process for shifting carbon monoxide to hydrogen mentioned earlier.

The point of discussing here in some detail our coke-agglomerating and fast fluidized beds is not so much to argue that these represent the only, inevitable line of development of new technologies for a Coalplex. Rather, it is to illustrate further a point made earlier, that the nation's expenditures upon coal engineering have been far too small to produce results commensurate with its needs. We have noted that groups working on SNG processes are just now beginning work on their proposed first coal-processing steps on a practical scale. Their too small budgets have postponed a confrontation with reality. Worse, when even they could not conduct experiments on a large scale, what chance was there for work on novelties which could only be studied on such a scale?

The fast fluidized bed cannot be studied on a small scale, for example. In small equipment, instead of the relatively "smooth" behavior of a large bed, the recirculation of powder to the bottom of a bed operating at 10 feet per second would produce an irregular "slugging" phenomenon discouraging to the investigator. Lurgi Chemie und Huettentechnik GmbH, Frankfurt (West Germany), the developer of the fast fluidized bed, had to have the courage to spend a substantial sum on large-scale tests before commercialization was possible. It is noteworthy that, once the decision was made to go ahead, things moved quickly. About eight years elapsed between the first "model" tests at atmospheric temperature and commercialization of the fast fluidized bed in a profitable unit for calcining aluminum hydroxide to make cell-grade alumina. Lurgi appears now to have captured the world equipment market for this step in aluminum production, less than two years after the start-up of the first unit.

The parallel with Solihull's developments for SNG from petroleum is striking. Here, too, eight years encompassed the first experiments with oil and commercialization, which promptly led to a commanding position in the world market for process technology. The two main ingredients of Lurgi's and Solihull's success were willingness to experiment early on a practical scale and an aggressive team for sales and service to licensees. With these two ingredients, an idea can be carried to commercial success in roughly 6 to 10 years; without them, development can stretch on forever.

Inadequate support for work on coal gives an overwhelming advantage to ideas which appear to be "safe" extrapolations from experience. There is no money to spare for exploratory research along new lines of thought, or for critical experiments, if they be expensive, to prove out new thought.

This is not of course to argue that support should be withdrawn from work now in progress. Yet we must be concerned lest in our haste and in our desire to see something tangible quickly we merely build plants based upon current knowledge and a poor fusion of modern chemical engineering with the ornery properties of coal. Too much money and hope committed to such installations could make it difficult to fund work on more advanced technologies. We will have to be willing to gamble large sums, as the nuclear engineers have done so often, on large experiments if the Coalplex is to be ready in time.

STRATEGIES FOR CLOSING THE DEVELOPMENT GAP

Our theme has been the need for large-scale experiments on coal- and oil-treating procedures. Chemical engineers do not yet command sufficient science to design equipment for such procedures in absence of large-scale experience. The research literature on fluidization, for example, has mainly heuristic value. Lurgi could not use this literature to predict the behavior of a fast fluidized bed in advance of experiment. Now, like the Russians after Alamogordo, we know that "it can be done," but we must conduct experiments ourselves before we can catch up with this new German art. Although agglomerating fluidized beds now serve several industries, no scientific literature on such beds is available to help us evaluate opportunities for applying additional beds of this type, let alone design them.

The U.S. has the skills, but there is a problem in mobilizing talent for the needed experiments. Most men having direct knowledge of the problems of large-scale experimentation of this type are employed by petroleum and chemical companies and vendors catering to these industries. Few, perhaps none, are at work in gas or electricity companies, the industries having the most urgent problems. Not many work in companies supplying equipment to sellers of gas and electricity, and this raises problems because of the tendency of electricity executives to prefer dealing with a few trusted suppliers.

Because our needs are so urgent, we will have to find extraordinary arrangements for organizing the necessary talent. We will have to bring together men who understand large-scale experimentation; who know the commercial value of getting on with it promptly; who can identify early commercial opportunities and can plan development strategies to gain profits with which to support work on long-range or more difficult questions; and not least important, men who understand the commercialization of new processes, how to license and serve licensees, how to export technology, and how to conduct defensive development to maintain a position once it is achieved.

Sending a task force to Japan might be in order to study how MITI is able to put together the remarkable quadrumvirate of government, industry, universities, and banks which is apparently responsible for much of Japan's success in process development and licensing. This is still a mystery here. Visits to Solihull, Institute de Francais du Petrole, the group of Lurgi firms, and Technicon could also be useful.

Since private firms have lost their former zest for risky process development, it is tempting to suggest the creation of patent pools bringing the best available ideas to bear upon the chosen goals, as MITI has done in Japan. Rewards to firms contributing patents and participating in the effort must be appropriate. Conn (1971) has called attention to the way in which the patent policies of some government agencies have narrowed their choice of organizations to do their work. Companies already holding proprietary positions in a field have not bid, and the government has had to select a bidder from among the less knowledgeable, whose primary motive may well have been to acquire expertise at government expense.

Perhaps it is not too late to form a pool of companies interested in procedures for direct desulfurization of residual oil, to convert this flagging development race into a cooperative effort toward a strong national position, rather than private, in this important technology. Japan has already mounted a large cooperative program of catalyst development toward the objective of 0.3% sulfur via direct desulfurization by 1985. Nippon Mining and Idemitsu Kosan have recently announced new catalysts to be tested commercially in Japan, which has more than 100,000 barrels per day of commercial direct desulfurization capacity; we have none.

Since commercial firms are reluctant to give up the services of senior men, it would be tempting to suggest the creation of an organization on the model of the Kellex Corporation, which was so successful in bringing together just exactly the right set of men to create the first gaseous diffusion plant on schedule. Perhaps a time limit might be set on the loan of a man to Kellex, or perhaps even on the existence of the organization itself, which would disband after a time judged sufficient for its emergency job.

Whatever the organization, the men should be assigned broad goals and given considerable freedom in selecting paths for study. For quick and sure results, we cannot afford the lumpiness in the effort which would follow from too heavy reliance upon the contract or firm or agency with a single narrow mission. It would be a great mistake to compartmentalize the effort among the goals—SNG vs. industrial gas vs. clean power gas and clean electricity vs. liquid fuels vs. low-sulfur coke for metallurgy vs. petrochemical feedstocks—or among the raw materials—pitch vs. coal vs. oil shale—or between environmental vs. commercial interests or between near-term vs. long-term objectives.

Key men should think regularly about *all* of the problems of the fossil fuel industries. Work on one feedstock cannot fail to help work on another, or one objective, another.

It might be argued that the support which the Office of Coal Research has given the Institute of Gas Technology has been counterproductive in at least one important respect: the energies of this small research institution were deflected from what was the earlier chance, process for SNG from petroleum feedstocks.

The importance of assigning near-term as well as distant goals depends upon more than just the desire for an early benefit. A development engineer, especially a young engineer, is seriously handicapped if he works for years under no great urgency to provide engineering designs for a full-scale plant which he knows will actually be built. There is a coziness in this circumstance hard to resist, which leads to the temptation to resort to dodges convenient for getting on with small-scale work but not suited for the field. A contract with but one long-term goal is apt to generate information having little or no value when the goal finally becomes immediate.

It is important to understand the fatal defects of the contract for a narrow mission. Men staffing the contract quickly become captives of their original idea. They cannot break away from it without losing their position. They are forced to exude an optimism which they may not feel. They might see failure coming, yet lack the flexibility to avoid it. Failure may become evident to the outsider only much later, and even then failure cannot be certified without a costly delay for investigative exercises calling upon appropriate authority. The men tend to enjoy freedom from criticism, even criticism which might be constructive, since outsiders are reluctant to say anything which might place the contract in jeopardy. Vested rights tend to arise in results obtained under the contract, for, in order that "duplication" may be avoided, other contractors, even those charged with a similar mission, are warned away.

The contract of narrow mission tends to create contract monitors in Washington, sometimes men with little experience, who have an even narrower term of reference than men working on the contract. This has led in at least a few cases to the ultimate absurdity of development work that proceeds little step by little step with release of funds for each step at the pleasure of a contract monitor far from the scene of activity.

A better role for government would lie at a higher level of decision. No doubt MITI participates fully in choosing objectives of Japan's development efforts, but it may be guessed that MITI has seasoned experts on its side of the table. From conversations with Japanese visitors, it may also be guessed that they keep environmental as well as commercial considerations well in mind.

PRIORITIES

In setting a priority for a technological development, the U.S. must ask, does the rest of the world want it? What will be its effect upon our balance of payments?

Over the long run, if our descendants are to enjoy something like our own industrial culture, they must rely upon energy sources which are essentially inexhaustible. Obviously, work on these ultimate energy sources must go on now if they are to be ready in time.

For many decades, however, technologies for conversion and utilization of fossil fuels will be a major and essential activity throughout the world. It would be folly to suppose that we can make a sudden shift from fossil fuels to an as-yet-undeveloped technology. The history of nuclear power illustrates this point. Its principles were demonstrated in December 1942 at Stagg Field. The atom first produced electricity in December 1951. Yet Hottel and Howard (1971) point out that nuclear power contributed only about $\frac{1}{4}$ as much energy in 1970 as the burning of wood. It is now coming on fast, but policy makers should note well the 20-odd year lag between its beginnings and its ability to contribute significant amounts of energy. This history can be expected to repeat itself for any of the new technologies. Because of the fossil fuel development gap, the U.S. is in immediate danger of becoming too reliant upon foreign sources of fuels. Dollars invested now to close this gap could be reflected many times over in an improved balance of payments in the 1980s, or sooner. Such can probably not be said of dollars invested today in the new technologies.

REFERENCES

Alpert, S. B., 1971. The future of fossil fuel technology. Paper presented at meeting of American Association for the Advancement of Science, Philadelphia, Pennsylvania, Dec. 28.

————, et al., 1971. Dealing with sulfur in residual fuel oil. In *Power generation and environmental change*, eds. D. A. Berkowitz and A. M. Squires. Cambridge, Mass.: MIT Press.

Ayrton, W. E. *On the economical use of gas engines for the production of electricity.* Translation of lecture delivered 28 September 1881 at Electrical Exhibition, Paris, France. London: W. Dawson & Sons (1881?).

Bienstock, D., et al., 1971. Environmental aspects of MHD power generation. *Proceedings of 1971 Intersociety Energy Conversion Engineering Conference* (held in Boston, Massachusetts, August 1971), pp. 1210–1217, Society of Automotive Engineers (New York).

Conn, A. L., 1971. Patent policy in government contracts, *Chem. Eng. Progr.* 67, no. 11: 31–32.

Dent, F. J., 1966. The Melchett Lecture for 1965: Experiences in gasification research. *J. Inst. Fuel* 39, no. 304: 194–207.

Elliott, M. A., 1961. Coal gasification for production of synthesis and pipeline gas. Paper presented at meeting of American Institute of Mining, Metallurgical, and Petroleum Engineers, St. Louis, Missouri, Feb. 26 to Mar. 2.

Graff, R. A., et al., 1971. Capturing sulfur with calcined dolomite. *Proceedings of Second International Clean Air Congress* (held in Washington, D.C., Dec. 1970), pp. 764–771. New York: Academic Press.

Hottell, H. C., and J. B. Howard, 1971. *New energy technology: Some facts and assessments.* Cambridge, Mass.: MIT Press.

Keith, P. C., 1971, Peapack, New Jersey 07977, personal communication, Sept.

Latta, N., 1910. *American producer gas practice.* New York: Van Nostrand.

Reh, L., 1971. Fluidized bed processing. *Chem. Eng. Progr.* 67, no. 2: 38–63.

Robson, F. L., 1971. United Aircraft Research Laboratories, East Hartford, Connecticut 06108, personal communication, Aug.

———, et al., 1970. Technological and economic feasibility of advanced power cycles and methods of producing nonpolluting fuels for utility power stations. Report from United Aircraft to National Air Pollution Control Administration, United States Department of Health, Education, and Welfare, Dec.

Robson, P. W., 1908. *Power gas producers: Their design and application.* London: Edward Arnold.

Ruth, L. A., et al., 1971. Desulfurization of fuels with half-calcined dolomite. Paper presented at Los Angeles meeting of American Chemical Society, Mar.; submitted to *Environ. Sci. Technol.*

Schroeder, W. C., 1962. Hydrogenation of Coal. U.S. Patent 3,030,297 (Apr. 17).

Squires, A. M., 1967. Cyclic use of calcined dolomite to desulfurize fuels undergoing gasification, *Advan. Chem. Ser.* 69: 205–229.

———, 1970. Clean power from coal. *Science,* 169: 821–828.

———, 1971. Clean power from coal, at a profit. In *Power generation and environmental change,* eds. D. A. Berkowitz and A. M. Squires. Cambridge, Mass.: MIT Press.

———, and R. Pfeffer, 1970. Panel bed filters for simultaneous removal of fly ash and sulfur dioxide: I. Introduction. *J. Air Pollution Control Assoc.* 20, no. 8: 534–538.

———, et al., 1971. Desulfurization of fuels with calcined dolomite: I. Introduction and first kinetic results. *Chem. Eng. Progr. Symp. Series* 67, no. 115: 23–34.

Sternberg, H. W., et al., 1970. On the solubilization of coal via reductive alkylation. *Amer. Chem. Soc. Div. Fuel Chem. Prepr.* 14, no. 1 (May): 1–11.

MICHAEL FORTUNE

Synthetic Fuels From Coal

Coal is now associated in the public mind with the worst that an energy-based technology has to offer, and rightly so. Yet, it is possible with today's chemical engineering knowledge to convert into synthetic substitutes for natural gas, power plant fuel, crude oil and gasoline. These substitutes can continue to supply gas and oil when the naturally occurring forms of these fuels become depleted. It is also possible to manufacture from coal a gas that can be burned cleanly in a "combined cycle" power plant at higher efficiency and with markedly less thermal and atmospheric pollution than is attainable in coal- or oil-fired electric plants. Furthermore, the technology of synthetic fuel from coal allows the consumption of both eastern coal, which would otherwise be unacceptable because of its high sulfur content, and western coal, which, due to its remoteness from the larger markets in the east, is too expensive to transport in solid form.

With technologies now under development, coal can be converted into essentially four types of synthetic fuel. One is synthetic pipeline gas, which could fulfill the functions of natural gas. A second promising type is a cheap special-purpose gas designed for high-efficiency power plants. A third is "refined coal," a low pollution fuel that can be used in liquid, chunk, or pulverized form in existing power plants, and which offers several advantages over coal and oil in the design and operation of new plants. The last type is synthetic petroleum distillates, including gasoline and low-sulfur fuel oil.

GASIFICATION OF COAL

Observe a wood fire in your fireplace closely sometime and you may see occasional jets of flame emanating from the hot wood, accompanied by hissing and crackling sounds. If you can obtain a few lumps of coal for burning, the effect will be even more striking: jets of yellow smoky gas will shoot out of the crackling coal a distance of several inches. In both cases, heat is driving off volatile gases from decomposing fuel in the interior of the chunks in the absence of air. The miniature gas works in your fireplace illustrate a phenomenon that was harnessed centuries ago and that may be coming of age again after a period of neglect. For, unlike the raw coal from which it is derived, synthetic gas can be easily cleaned and transported through pipelines to homes and community power plants.

The Rev. John Clayton was the first person to report collecting gas from coal. In 1688, he heated coal in a retort and noted that first "there came over only phlegm, afterwards a black oil and then, likewise, a spirit arose . . . the spirit which issued out caught fire . . . and continued burning with violence" (Stewart, 1958).

George Dixon built the first pilot to make gas about 1760, but its destruction by an explosion caused him to abandon the effort. Beginning in 1792, William Murdoch experimented with various gasifiers and with piping the gas to fixed street lights. By 1807, the first commercial coal-gas plant was lighting factories and homes in Manchester. In 1812, the first gas utility, the Chartered Gas Light and Coke Company, was organized to supply gaslight to London.

Shortly afterwards, gas-purifying equipment using lime slurries came into use. F. A. Winsor, the founder of the utility industry, strongly advocated his gas as a clean alternative to coal and tallow oil.

The method employed in these earlier gas-works is properly termed carbonization or coking, which leaves a solid residue (coke) of high carbon content. By 1824, "water-gas" was manufactured by passing steam through a bed of white-hot coke. Water-gas, which is half-and-half carbon monoxide and hydrogen, has a lower heating value (300 British Thermal Units per Cubic Foot, or BTU/CF) than coal-gas, but it augments the energy in the gas per unit of coal. (Coal-gas driven off during coking has a heating value of about 500 BTU/CF, compared to natural gas, 1000 BTU/CF.) The poor heating value of water-gas may be bolstered by adding some rich gas (1700 BTU/CF) from the cracking of oil on hot bricks. Gilliard, in 1849, introduced the first intermittent partial-oxidation process, wherein a blast of air burned part of the coke to generate the heat needed for the gas-generating reactions. Air and steam were alternately passed through the oven as its temperature was raised and lowered respectively.

By 1885, plants were built that continuously and completely gasified the coal. They employed a two-stage process remarkably similar to many "modern" projects now in the R&D stage. Carbonization and coke gasification processes were situated close to each other, so that the mutual exchange of coke and gas conserved much heat.

By 1933, the Lurgi process, the only early method still under serious consideration today for further development, was introduced. In the Lurgi process, coal is gasified with oxygen and steam under high pressure. This technique combines high yield and efficiency—70%—and a product with reasonably high heating value of the gas—475 BTU/CF. (The efficiency of a gasification process is defined as the heating value of the product gas divided by the heating value of the raw coal entering the plant.) The Lurgi process has since provided gas for England, Germany and South Africa. Lurgi's potential is illustrated by a new German power plant burning synthetic gas from this process in a combination of gas and steam turbines. Not only does such a power plant deserve credit for its low thermal and air pollution, but combined capital costs for both the gasifier and the electric plant are remarkably low—equivalent to $90 per kilowatt in German funds.

Synthetic Gas in the United States

In this country, demand for synthetic gas declined first when Edison's lamp gave better lighting, and later, after WW II, when pipelines were first built to carry gas from southern fields to the industrial centers of the Northeast. These pipelines opened up vast new markets for natural gas, which, though previously flared at the oil well, was now in great demand as a clean, cheap fuel. It replaced much coal and fuel oil as well as synthetic gas as a source of heat. Today, interest in synthetic gas is growing again, owing to the limited extent of our natural gas supplies.

In the last five years, research on gasifying coal has been intensifying. In his message on energy of June 4, 1971, President Nixon promised to double the funding for pilot plants under the Office of Coal Research (OCR) to $20 million. Later, the Administration announced that the budget for that Office would be increased by $3 million to initiate work on the generation of electricity from low-BTU gas from coal.

Now, commercial synthetic gas plants will again be built in the United States. The El Paso Natural Gas Company has announced (Coal News, 1971) that it will build a coal gasification plant in New Mexico based on the Lurgi process. It will cost more than $400 million and will provide 250 million cubic feet of pipeline-quality gas per day after 1976. Commonwealth Edison Company of Chicago will build an experimental plant, also based upon the Lurgi process, to supply low-BTU gas to a 125 megawatt boiler for its Powerton Station, at a cost of $17 million.

There are two quite different synthetic gases which will be made for different uses. The first is a substitute for natural gas. Known as pipeline gas, or rich gas, it must have a high heating value (about 1000 BTU/CF), must consist primarily of methane or higher hydrocarbons, and a little hydrogen, and should not contain any poisonous compounds. The second is a gaseous fuel for electric utilities. Variously termed producer gas, power gas, or merely low-BTU gas, this fuel possesses lower heating values (generally around 125-175 BTU/CF) and may also have some toxic properties. It is composed of hydrogen, carbon monoxide, a little methane and inert gases.

According to both the Lurgi art and the methodology used in Interior Department coal research, pipeline gas would be synthesized in part by distilling methane directly from coal, and in part by synthesizing methane from water-gas.

It may be produced at the mines because the cost of transporting gas through a pipeline averages two-thirds of the cost of hauling an equivalent amount of coal by shuttle train (OCR, 1970b). (However, transporting gas by pipeline is *more* expensive than hauling an equivalent amount of coal by a *unit train* that is specialized to carry nothing else but coal.) Although the expense of the necessary clean-up and methanation steps is considerable, most of the coal research in the U.S. is aimed at pipeline-quality gas rather than low-BTU gas suitable

Figure 1   IGT hygas process—steam-oxygen gasification

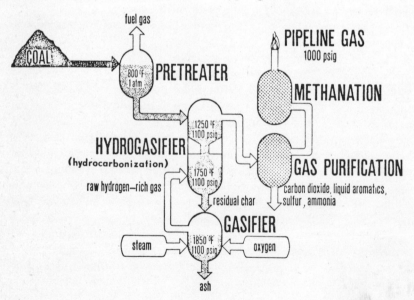

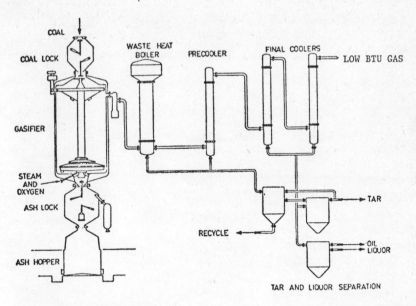

Figure 2 Lurgi Gasifier

only for generating electricity. (See Figure 1 for a diagram of a sample pipeline-gas process.)

Because low-BTU producer gas is expected to be manufactured at the site at which it is to be consumed, that is, primarily at electric power plants, and because there are fewer stages in the process, its cost will be less per BTU than that of pipeline gas. (See Figure 2 for a diagram of the first step of a sample low-BTU gas process.)

THEORY OF COAL GASIFICATION

Coal is a complex substance with a molecular weight in the natural state on the order of 3000. When heated in the absence of air, it gives off gases and liquids (collectively called volatiles) of low molecular weight, and leaves a solid residue (char) of greater molecular weight and lower chemical reactivity than the coal. Because the energy-rich volatiles are fairly easily driven off from coal, all gasification processes attempt to maximize the amount of these gases that are released. The processes differ in their approach to gasifying the char, where most of the energy content of the original coal remains.

Table 1 lists the important chemical reactions occurring during char gasification. These fall into three groups: production of water-gas, formation of methane, and oxidations.

Table 1 Chemical Reactions Important to Gasification

Reaction	Does it Absorb or Release Heat?	Heat of Reaction Kcal./mole	Name
Production of Water-Gas:			
(1) $C+H_2O\rightarrow CO+H_2$	endothermic	$+31.38^a$	Water-gas
(2) $C+2H_2O\rightarrow CO_2+2H_2$	endothermic	$+21.24^a$	
(3) $CO+H_2O\rightarrow CO_2+H_2$	endothermic	$+\ 0.45$	Shift Conversion
(4) $C+CO_2\rightarrow 2CO$	endothermic	$+41.53$	
Production of Methane:			
(5) $C+2H_2\rightarrow CH_4$	exothermic	-19.1	Hydrocar-bonization
(6) $2CO+2H_2\rightarrow CH_4+CO_2$	exothermic	-60.64	
(7) $CO+3H_2\rightarrow CH_4+H_2O$	exothermic	-49.47	Methana-tion
(8) $CO_2+4H_2\rightarrow CH_4+2H_2O$	exothermic	-38.32	
Oxidations:			
(9) $C+1/2O_2\rightarrow CO$	exothermic	-26.55	
(10) $C+O_2\rightarrow CO_2$	exothermic	-94.4	
(11) $CH_4+2O_2\rightarrow CO_2+2H_2O$	exothermic	-190.9	

[a]This value reflects the fact that water is in gaseous form in the reaction. For all other reactions, water is assumed to be in the liquid phase.

Two principles of physical chemistry are helpful in interpreting Table 1:

a) Normally, simple one-step chemical reactions proceed more quickly at higher temperatures.
b) Among reactions that proceed to equilibrium, an increase in temperature will favor the more endothermic (heat-absorbing) process, but will suppress the exothermic (heat-releasing) process.

Reaction (1), the historic "water-gas" reaction between steam and coal, will, in general, be the main means for converting char into gaseous form. Reaction (2) also occurs, but is usually minor compared to (1).

Following principle b), water-gas reactions require a rather high temperature, above 1650°F, for suitable conversion rates. In fact, the amount of carbon that is converted decreases drastically below 1560°F (850°C). In the original high-pressure Lurgi reactors, the bed temperature was limited by the softening point of the ash, typically 2010°F (1100°C). Recent experiments in England in handling liquid ash (slag)

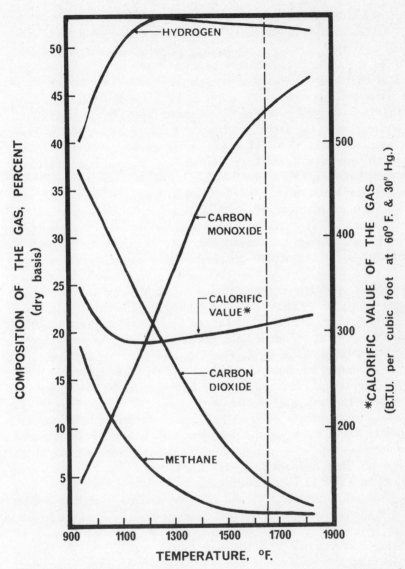

Figure 3 Variation with temperature; composition and calorific value of water gas made from steam and carbon under equilibrium conditions (approximations from equilibrium data)

may permit future gasifiers to be operated at very high, favorable temperatures (see Fig. 3).

In light of principle b), it is important that water-gas reactions are endothermic, i.e., that they absorb heat. The heat of reaction is nearly always provided by burning either raw coal or char at some point in the plant. If a portion of the batch of coal undergoing gasification were burned to supply the necessary energy directly, then Reaction (10) would be the desired combustion because it releases more heat per pound of coal and requires less oxygen per Kcal than other combustions. However, because Reaction (9) is more endothermic (less exothermic), it predominates at the high temperatures typical of gasifiers.

If a clean, low-BTU power gas is desired, it can be formed by gasifying char, as in Reactions (1) and (2) provided with a source of heat and a clean-up stage. In the Lurgi gasifier (Fig. 2), coal is devolatilized in the upper portion of the vessel as hot water-gas flows upwards past it. As the char sinks towards the bottom, it is gasified with steam and air—the heat of reaction being supplied by partial oxidation of the char with air. The synthesis of high-BTU pipeline gas, which must be essentially methane, on the other hand, is considerably more complex, hence more expensive. Three approaches to pipeline gas have been explored.

In the first approach, the maximum yield of methane is achieved by exposing raw coal to relatively pure hydrogen at high pressure in a hydrocarbonization step. Workers at the University of West Virginia, who have made an economic assessment of all pipeline-gas schemes under consideration (Wen et al., 1972), believe that this approach (the Hydrane process of the Bureau of Mines) will give the highest efficiency and the cheapest gas. Unfortunately, it has received the least attention and the least funding.

In the second approach (Fig. 1), coal is contacted at high pressure with water gas to drive off the volatiles; then if the contact is maintained long enough, the carbon in the char combines directly with hydrogen to yield methane in a step known as hydrogasification (reaction [5] in Table 1). Being exothermic, this reaction reduces the need for the oxidation reactions (9) and (10) by yielding heat that can be applied to the formation of water-gas (reaction [1]). Most of the hydrogen for hydrogasification is formed in the separate gasification of residual char withdrawn from the main vessel.

In the third approach, water-gas is generated from char, is purified, partially undergoes a "shift conversion" (reaction [3]), and then is "methanated" (reaction [7]). In the attempt to maximize the yield of methane, this approach results in certain trade-offs. A greater portion of the coal must be oxidized to provide heat than is required by the other two approaches. Also, some undesirable CO_2 is formed 1) in partial oxidation of the coal, 2) in "shifting" the water-gas to the proper 3:1 hydrogen:carbon monoxide ratio according to reaction (3), and 3) in the unavoidable reaction (6).

In practice, all of the above three processes occur at the same time, to a greater or lesser extent, depending on design. The efficiency of the various schemes tends to be higher with a greater extent of hydrogasification and a lesser need to "methanate" water-gas. Typical efficiencies for pipeline-gas plants range from 55–75%. Power-gas generators, not troubled by the constraint of maximizing methane yield, can achieve more acceptable over-all efficiencies of 70–90%. As mentioned earlier, one reason for the inefficiency is the necessity to burn a portion of the coal to get the energy for the endothermic process. Heat losses from radiation, stack gases, water vaporization and even cooling water make these processes less than 100% efficient.

In the manufacture of pipeline gas, air cannot be employed in combustion because inert nitrogen, which comprises four-fifths of air, would so dilute the water-gas that its heat value could not be upgraded to pipeline quality. However, since oxygen manufacturing plants represent a significant expense for a high-BTU gas works, it pays to skimp on pure oxygen wherever possible.

The need for an oxygen plant would be obviated if the combustion and the gasification occurred in separate chambers, and a cycling flow of some material would transfer heat from one chamber to the other. Several processes use materials such as dolomite (the CO_2-acceptor process) or molten sodium carbonate (Kellogg), or electric power (HYGAS) to transfer energy from the combustion chamber to the gasifier.

Keep in mind that this is an energy-robbing process. Just to convert the fixed carbon to a gas according to the water-gas reaction requires the energy from combustion of up to 30% of the raw coal. The percentage will always be less than this, of course, because the volatile fraction of the coal will readily form energy-rich hydrocarbons. The wide variations in volatile content of different coals may require development of schemes tailor-made for each grade of coal. Energy is also required to maintain the reactor at the high 900°C+ temperature, to raise steam, to preheat the coal and air, to pressurize the system, and to operate a maze of auxiliary equipment. If pipeline gas is to be made, energy is needed for the oxygen plant and/or hydrogen generators, and some energy is lost in carrying out the shift reaction (3).

Figure 4 shows that the yield of methane and the heating value of the product gas increase rapidly with pressure. Unfortunately, the CO_2 fraction also rises. In fact, the fraction of CO_2 rises as rapidly as the fraction of either hydrogen or carbon monoxide falls; this shows that the latter gases are reacting as $2H_2 + 2CO \rightarrow CH_4 + CO_2$. Other advantages of pressure are a smaller gasifier vessel, cheaper clean-up and methanation equipment, and less compressor work. For these reasons,

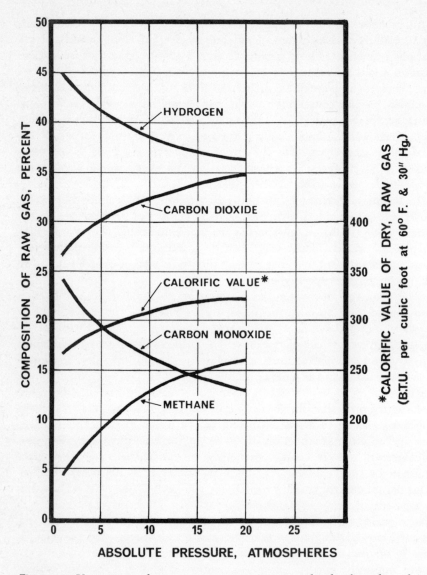

Figure 4   Variation with pressure in composition and calorific value of raw gas as made in practice from brown coal, employing oxygen and steam as gas-making fluids

all the modern existing or contemplated gas works operate under pressure.

Because reactions (5) through (10) in Table 1 are exothermic, methane will be less readily formed at high temperatures (see Table 1 and Fig.

3). The components of water-gas, H_2 and CO, behave in an opposite fashion. The calorific-value curve (Fig. 3) shows a minimum that separates the desirable temperature regions for hydrogasification and for synthesis gas manufacture.

The volatile fraction of the coal is associated with a nuisance, *agglomeration*. This is a property that some coals exhibit in a closed container; they swell, become plastic, and fuse into a massive glob. A superficial oxidation of the coal surface in a blast of hot air will destroy this property, but such a procedure wastes much of the valuable volatiles. Some processes obviate this pretreatment by carefully keeping the temperature just below the plastic point of the coal at each stage, but, as might be expected, this is expensive. Eastern U.S. bituminous coals, those closest to the markets, tend to agglomerate the most; whereas, those from the western U.S., with little tendency to agglomerate, can usually be gasified readily. The prevalence of low-caking coals in Europe may explain why they have been more successful in marketing synthetic town gas.

TYPES OF PLANT FOR GASIFYING COAL

The vessel in which coal is gasified is a vertical pressurized cylinder, from 3 to 15 feet in diameter and often over 100 feet tall. There are usually several zones, one above the other, where different reactions occur. The placement of these zones and the nature of the fuel bed differ among the various processes.

The oldest arrangement is the *fixed bed* (or, more properly, *gravitating bed*), wherein a layer of coal a few feet thick lies on a grate. Ash is removed in a dry state from below the grate. Because the surface area of the coal available for reaction is small, the rates of gas production tend to be low. The temperature must be limited to prevent agglomeration and melting of the ash. More recent experience in handling liquid ash (at temperatures around 2500°F) has opened up the possibilities of *slagging* beds, which allow elevated temperatures at which the rapid formation of water-gas is highly favored.

In *entrained* and *concurrent* gasifiers, the coal is pulverized and swept through the vessel in the gas stream. The fuel converts to a gas very rapidly without the formation of tar, but the thermal efficiency is low unless expensive heat-recovery equipment is installed. The unreacted char must be separated from the gas leaving the reactor and be re-injected.

A *fluidized* bed consists of small lumps of fuel that are lifted bodily and kept in turbulent suspension by a rapid upward flow of gas. The weight of the fuel lumps is counterbalanced by the drag of the gas stream past them. The heat transfer and the reaction rates in fluidized beds are remarkably rapid; however, small particles (fines) of carbon are swept away in the gas stream.

Although they are indeed the heart of a gas-manufacturing plant, the gasifiers require a maze of auxiliary equipment. Coal storage and preparation require a great deal of space. It may be necessary to pre-treat the coal to destroy its caking properties. Conventional boilers raise steam for the water-gas reaction and for numerous other stages. Scrubbers remove the acid gases from the product and channel the noxious hydrogen sulfide to a Claus plant where it is converted into sulfur. The hot ash must be safely withdrawn from the pressurized vessel and dumped or sold. Just before the product gas enters the pipelines, it is dried with glycol.

Many other facilities are needed to support a pipeline-gas plant. Oxygen is manufactured by liquifying air and distilling it. A hydro-carbonization process requires a separate gasifier to produce hydrogen from residual char and water. The shift converters adjust the ratio of hydrogen to carbon monoxide to the proper 3:1 value for the next step, methanation, wherein a catalyst assists in transforming synthesis gas to methane. Pipeline gas must also be monitored constantly so that its quality remains high.

ENVIRONMENTAL ASPECTS OF GASIFICATION

The National Environmental Policy Act of 1969, section 102 (2)(C), requires all federal agencies to file environmental impact statements of their projects. Statements on the impact of existing and proposed pilot plants should help us appraise the effects. Unfortunately, neither the Office of Coal Research nor the Bureau of Mines (both in the Department of Interior) has published such statements on their programs in coal research. At the very least, the law directs them to write an impact statement for each of the eight or so existing and proposed pilot plants.[1] Experience at these experimental plants will assist in formulating a general statement on the environmental impact of coal gasification. In the following analysis, a proposed commercial BIGAS plant is analyzed for its environmental impact. (Refer to Table 2 for specification of the plant.)

[1a.] HYGAS; Chicago, Illinois.
[b.] Consol CO_2 Acceptor; Rapid City, South Dakota.
[c.] Char Oil Energy Development; Princeton, New Jersey.
[d.] Project Gasoline; Cresap, West Virginia.
[e.] BIGAS; Homer City, Pennsylvania.
[f.] Solvent Refined Coal; Tacoma, Washington.
[g.] Bureau of Mines Low-BTU Coal-gas; Morgantown, West Virginia.
[h.] SYNTHANE; Pittsburgh, Pennsylvania.

Table 2   Specifications for a Proposed Commercial BIGAS Plant

Capacity—250,000,000 standard cubic feet of synthetic pipeline gas per day.
Coal Requirements—605 tons per hour (as received; excludes storage).
Water Requirements—Cooling water 262,580 gallons per minute, process water
  5,790,000 pounds per hour.
Demand for Electric Power—41,860 KW generated on site.
Sulfur By-product Produced—14.7 tons per hour.
Slagged Ash Disposed—11.4 tons per hour.
Operating Pressure—80 atmospheres.

Source: U.S.O.C.R. R&D Report #60, February 1971.

AIR POLLUTION

Potentially, the most serious effects of the gasification plant itself are air pollutants. These fall into two categories: the fumes normally resultant from the combustion of coal, and the sulfurous compounds released from the gas clean-up stages. About 16% of the coal entering the plant is diverted to boilers and superheaters that raise steam for the water-gas reactions, shift reactions, on-site electric power generation, oxygen compressors, and wherever a little heat is needed. A large amount of electricity is needed (42 megawatts for the proposed BIGAS plant), mostly to pulverize and otherwise prepare the coal; the rest is used for pumping. All the steam would be raised in traditional coal boilers, one of whose traditions is air pollution. About 170 tons per day of fly ash would have to be removed by filters or precipitators, and a comparable quantity of sulfur dioxide must be dealt with.

This problem would be completely eliminated if a portion of the product gas were burned to raise steam, although, apparently, no projected plant incorporates this idea—probably, because of a slightly lower efficiency and the need for auxiliary fuel to start up the system. Since it is wasteful to use enriched gas for process heat and power needs, full scale plants might be designed to bleed appropriate amounts of low-BTU gas from the system or have separate power gas production facilities.

The greater and less predictable problem arises from the inability to contain all the sulfurous gases that are removed from the product gas. Pipeline gas must be exceptionally clean, not only because impure gas would annoy the housewife with a gas appliance, but because any trace of hydrogen sulfide or carbon dioxide would corrode the pipelines (in the presence of moisture). Producer gas destined for electric power plants must also be cleaned, as hydrogen sulfide will badly corrode the blades of a gas turbine, besides polluting the air with SO_2 after combustion. However, as the product gas is rendered cleaner and cleaner, it becomes more difficult to recover all the impurities as a by-product in the form of elemental sulfur, at least with existing technology. Either the cost of treating the emissions rises sharply, or the quantity of sulfurous gases vented to the atmosphere at the gasification site becomes greater.

The most common proposed clean-up system is the hot potassium carbonate scrubber. A solution of potassium carbonate in water absorbs the acid gases (CO_2, H_2S, and COS—Carbonyl sulfide) from the raw synthesis gas, is pumped to a regenerator, and there releases these acid gases at a higher concentration whereupon they can be treated. If the scrubber removes the acid gases in one step, then the hydrogen sulfide will be subsequently released at a concentration of 3.4% in carbon dioxide. This level is far below the minimum 15% needed to feed a conventional Claus plant which converts gaseous H_2S into solid sulfur. Since potassium carbonate preferentially absorbs H_2S over CO_2, H_2S could be removed in one stage prior to CO_2 removal in a later stage. If this is done, the first-stage regenerator will reject H_2S at a concentration high enough for treatment. The first stage removes 90% of the H_2S, but little of the carbon dioxide; the H_2S is released at a 15% concentration, high enough for the Clause plant to recover according to the reaction $2 H_2S + O_2 \rightarrow 2H_2O + 2S$. Historically, the Claus plant itself has not been completely leakproof; about 3–7% of the sulfur compounds are vented. A number of firms now offer systems to suppress emissions of sulfur from Claus systems. This is an easy task by comparison with suppression of SO_2 from the stack of a conventional electricity station.

The task of the second stage is to remove carbon dioxide and most of the remaining 10% of the hydrogen sulfide. (In a producer gas generating plant, where it would not be necessary to remove carbon dioxide, but would be best to capture all of the H_2S, the second stage will still be needed.) However, unless this hydrogen sulfide is to be subsequently vented to the air at the unusually high concentration of 3400 parts per million (ppm), a "molecular sieve," or some other process, would be needed to capture it and channel it to the Claus plant with an attendant increase in the gas price.

The third stage, the "sulfur guard reactor," removes all remaining traces of sulfurous gases from the gas which is to be sold on beds of ZnO granules. Eventually, the granules become ZnS and are discarded as solid waste. In a full scale plant, the zinc could be recovered.

An alternative to the above conventional gas clean-up scheme supposedly surmounts all the above difficulties. The Rectisol refrigerated methanol absorption process reportedly achieves "substantially complete" removal of hydrogen sulfide, increases the overall thermal efficiency by 3 percentage points, and cuts the price of product gas by

2.9¢ per million BTU. It loses significant amounts of methane, however.

The gasification schemes which utilize a cycling heat carrier remove substantial portions of the fuel's sulfur right in the gasifier. The cycling dolomite in the CO_2-Acceptor Process reacts with H_2S as soon as the gas emerges from the coal particles, carries it to the regenerator and then releases it as H_2S and elemental sulfur (OCR, 1970b). The SO_2 would be harder to contain than the H_2S and may contribute to air pollution. A dolomite acceptor process may also reclaim the sulfur as H_2S, which is easily reduced to elemental sulfur.

The conditions inside a gasifying vessel cause the sulfur in the coal to emerge as hydrogen sulfide rather than sulfur dioxide as in a conventional power plant. If there are routine emissions of H_2S, or any accidents that lead to large leaks, the hydrogen sulfide will be found to be more nauseous and noxious than the sulfur dioxide from an equivalent amount of coal.

In short, conventional air pollution clean-up technology involves a trade-off between purity of the product gas, price of the gas, and cleanliness of the air near the gasification complex.

OTHER ENVIRONMENTAL COSTS

Most gasification schemes will generate ammonia, which might be discharged as a water pollutant. The Char Oil Energy Development (COED) scheme produces, as about 7% of its yield, phenolic liquors that are unlikely to be totally reclaimed. A gasification plant will require extensive water-treating facilities.

Viewed in the context of total national energy consumption, another important environmental effect of the synthetic fuels is thermal waste discharges. Gasification of coal may or may not reduce the percentage of heat wasted through consumption of energy. The gas works do use a significant fraction of the heat content of the raw coal in the process of converting it, ranging from 10% to 45%. If the product substitutes for natural gas, nation-wide thermal waste will certainly increase. The fuel requirements for a synthetic pipeline gas market would thus average $1\frac{1}{2}$ times more than the requirements for a market consuming only natural gas, the extra fuel being burned at the gas works.

The thermal waste of a producer gas generator (which tends to conserve more energy than a pipeline gas generator) *can* be offset, but not necessarily by the greater efficiency of a combined cycle power plant relative to conventional plants. In a combined cycle, hot exhaust from gas turbines is used to raise steam for steam turbines. When the gas-purification stage operates at high temperatures, the efficiency of conversion of coal to low-BTU gas can exceed 90%, if the sensible heat of the hot clean gas is accounted for, as well as the potential heat that remains to be recovered by further combustion. Assuming extreme efficiencies, 90% for the gasifier and 60% for the combined cycle, the overall utilization of energy will be a respectable 54%. But, with efficiencies attainable with today's technology (75% for a Lurgi gasifier, 47% for the combined cycle), the overall efficiency quickly drops to 35%, slightly less than that of new conventional power plants. Thermal waste would be nearly halved in the extreme case, and increased about 1.2 times in the realistic case. Of course, if synthetic gas is supplied to an ordinary steam power plant in place of natural gas or oil, heat loss will be greater, just as in the case of pipeline gas.

The designs for gasification plants incorporate cooling towers. The nominally sized, 250 million cublic feet per day BIGAS plant, for example, will have two mechanical-draft cooling towers through which circulate 263,000 gallons of water per minute. The bulk of the cooling load resides in the oxygen plant, acid gas removers, and the turbo-generator condenser. The evaporation rate from the towers is 2626 tons of water per hour.

REFINED COAL: A SYNTHETIC SOLID OR LIQUID FUEL

A new product, "refined coal," may be able to economically supply clean fuel to existing equipment now fired with coal or high-sulfur fuel oil. Some think that refined coal will be more expensive than low-BTU gas for two reasons: it must be made in equipment operating at far higher pressure and the efficiency of its manufacture will be less (about 65% to 70% versus, perhaps, 90%). However, refined coal can be shipped and can probably be made available to the small user of fuel at a lower cost than any form of synthetic gas.

Refined coal is a solid at normal temperatures, but melts at about 150°C. It can be shipped either as a solid or as a hot liquid. The process can also be designed a bit differently to yield a think liquid at room temperature. By removing noncombustible portions of coal, the 10% to about 0.1% and can reduce the sulfur content to 0.8%, in one variation of the process, and as low as 0.2% in two other variations. The heat content of refined coal is higher than the original coal (15,900 BTU per pound versus a typical 12,500 BTU per pound of bituminous coal), and transportation costs are correspondingly reduced. If refined coal is shipped in a liquid state, once in storage tanks, it need only be pumped from the tanks and atomized for burning in a boiler, obviating the costs of stockpiling coal, recovering it from the pile, crushing it, pulverizing, and entraining in air for supply to a pulverized-fuel boiler.

A catch in the picture is the probability that nitrogen oxide emissions from the firing of refined coal will be unacceptably high. The refining process has the unfortunate property of concentrating all of the nitrogen in the coal into the refined product. It is well known that fuel nitrogen is converted, to a large degree, into nitrogen oxides, whatever the mode of combustion (surprisingly, even in the low temperature combustion that occurs in the fluidized bed boilers under development). No cheap means for removing nitrogen from refined coal has been suggested.

Three processes for refined coal have been proposed:

1) The Solvent-Refined Coal (SR-Coal) process of Pittsburgh and Midway Company, a subsidiary of Gulf Oil Company, developed with support from O.C.R. (OCR, 1970a).
2) A modification of the H-Coal Process of Hydrocarbon Research, Inc., developed with private money (OCR, 1968).
3) A process under study at the Pittsburgh Energy Research Center of the Bureau of Mines (Akhtar et al., 1971).

Products of each of the three processes are substantially identical, with the exception that SR-Coal contains far more sulfur. Figures 5, 6, and 7 are diagrams of the three processes, respectively. The processes, in their commercial embodiments, will differ in the operating pressure and the reactor in which coal is dissolved and reacted with hydrogen.

In the process, coal would be slurried with a highly aromatic, recycled oil, subjected to high pressure (at least 1000 pounds per square inch) and heated. At about 250°C, the coal begins to dissolve in the solvent. It is then introduced into a reactor with some hydrogen, which is generated elsewhere in the process from gases and naptha. The SR-Coal reactor is simply an empty vessel, while the H-Coal reactor contains particles of a cobalt-molybdenum catalyst buoyed by the rising flows of liquid and hydrogen, the catalyst forming an "ebullated bed." The Bureau's reactor is a long, coiled pipe filled with catalyst pellets through which the liquid slurry and hydrogen flow at velocities appreciably higher than in the other reactors.

Typically, about 90% of the coal would react to form a liquid that can be filtered from unreacted carbon and ash matter. The solid portion of the remaining 10% would, presumably, have to be discarded as waste. Solvent and light oils would be distilled from the filtered liquid, and the refined coal product would remain as a residue.

The two processes that use a catalyst can remove about 90% of the organic sulfur present in the raw coal. Without a catalyst, the SR-Coal process can remove about 60% of the organic sulfur. Table 3 gives a typical analysis of SR-Coal. Each process effectively removes all the pyritic sulfur. The Bureau's process can achieve about 0.15% sulfur content in the product, H-Coal can achieve 0.25%, while SR-

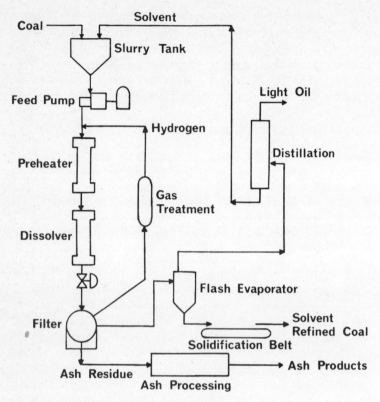

Figure 5 Solvent refined coal process

Table 3 Analysis of Solvent Refined and Raw Coal

	Contained in Raw Coal	Contained in Solvent-Refined Coal
Ash	6.91%	0.14%
Sulfur	3.27%	0.95%
Carbon	71.3%	89.2%
Volatile Matter	4.4%	5.1%
Oxygen	12.3%	4.4%
Nitrogen	0.94%	1.3%
Heating Value	28.0 million BTU per ton	31.9 million BTU per ton

Source: U.S. Office of Coal Research, #53.

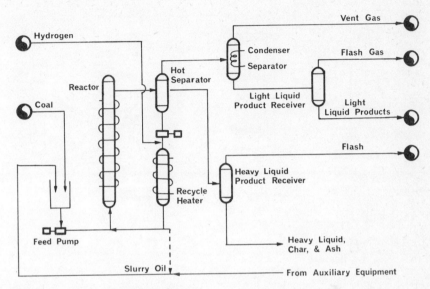

Figure 6   H-Coal bench scale unit

Coal appears limited to about 0.8%. Even the SR-Coal process depends somewhat upon the naturally catalytic nature of most coal ash, and the process does not work well for some coals where natural catalysts are not present in the ash.

At least one person prominent in the coal conversion field, Arthur M. Squires, believes that the success of the coal-refining process will not be demonstrated until equipment of large capacity becomes available, because an unpredictable, exothermic chemical reaction occurs soon after the coal dissolves (personal communication).

The Office of Coal Research and Consolidation Coal Company had, until recently, operated a program to demonstrate the conversion of coal to gasoline under the name "Project Gasoline." The National Academy of Engineering (OCR, 1970c), among others, investigated why the project's pilot plant at Cresap, West Virginia, was forced to shut down indefinitely in April 1970, after $20 million had been spent. Incompetent management, labor troubles, and numerous mechanical and process failures all conspired to prevent the plant from operating sufficiently long to reach the steady state conditions necessary to provide its own recycle solvent. It would have cost $3 million merely to renovate the plant.

During the past two years, a number of plans for revising the Cresap facility have been studied. At one time, it appeared that Cresap might be rebuilt to provide a test of the H-Coal process. Currently, it

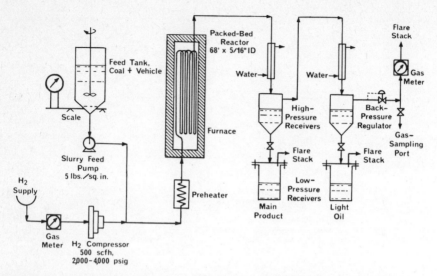

Figure 7 Simplified schematic diagram of hydrosulfurization pilot plant

appears that the Cresap process will be revised to provide a fourth approach to production of "refined coal."

IMPACT OF SYNTHETIC FUELS ON COAL RESOURCES

The question of depletion of coal resources is closely tied both to the substitution of coal for other fuels and to the problem of thermal waste because of the necessity to burn some fuel at less than 100% efficiency in order to drive the chemical reactions. Assuming a 25% heat loss at the conversion plant, if gasified and liquified coal were to replace all of today's demand for natural gas, oil, and coal, then the extraction of coal will be more than sextupled! If only natural gas and coal are substituted, then the coal industry must grow by a factor of three and one-half.

Strip mining will most likely accelerate, even if its present 45% share of coal production holds steady. There are indications, though, that wide-scale gasification will encourage relatively greater strip mining.

The untouched reserves of western coal, which represent most of the U.S.'s strippable coal reserves, may be tapped using either unit trains or by transporting synthetic gas or liquids in pipelines.

Western lands tend to be less biologically productive and less economically valuable than the rich farm land overlying the Illinois and Missouri coal fields and the (formerly) thick forests over the Appalachian coal. It is conceivable, then, that there will be less opposition to strip mining of western coal. However, continued opposition to strip mining on land in the Four Corners region of the west indicates that it is a source of disgust wherever it is practiced.

ECONOMIC CONSIDERATIONS

Rapid inflation, especially in construction and in the fuels market, and differences in accounting procedures make any comparison between cost estimates made at different dates meaningless. Furthermore, the great disparity between the cost per BTU of different fuels clouds the issue. No attempt is made here to compare the cost advantages of competing synthetic fuel processes; instead, only a rough comparison of the cost of different types of synthetic fuel with the cost of natural gas is given.

In mid-1972, the average price of gas paid by pipeline companies to producers was 21¢ per million BTU (weekly FPC News Reports) equivalent on an energy basis to a mine mouth price of about $5.25 per ton of bituminous coal. The cost of low-BTU gas suitable for generation of electricity is about twice to three times this figure, while the cost of synthetic pipeline-quality gas varies upward from four times the cost of natural gas. The lowest estimates are for processes that utilize lignite, a very cheap coal. The cost of synthetic liquid and solid fuels derived from coal seems to fall above about 70¢ per million BTU.

SUMMARY

There is no question that it is technically feasible to convert coal into practically any type of fuel desired. Low-BTU and moderate-BTU coal gas has been manufactured for a century and a half. With the Lurgi process, production of either high-BTU pipeline gas or low-BTU gas appears assured. Processes for pipeline gas that are believed to be better than the Lurgi process are now in the pilot-plant stage of their development. While it is difficult to synthesize gasoline, the production of low-sulfur fuel oil and low-sulfur char from coal is technically feasible.

The economic competitiveness of synthetic fuels with their natural counterparts is not yet a matter of fact. They are generally more expensive than oil, coal, and natural gas, but this may not be so in the future because the costs of coal and natural gas are rising dramatically. The generation of electric power with low-BTU synthetic gas is believed by some to be competitive, and it will probably account for the first wide-spread use of gasified coal.

Low-BTU synthetic gas is *desirable* because electric power plants fueled with such a gas will contribute markedly less to air and thermal pollution compared to today's plants. The type of power plant suited to burn low-BTU gas is the "combined gas and steam turbine" arrangement, in which the very hot flue gases (2400°F) from combustion expand through a gas turbine before they heat up the boiler in the conventional steam-turbine cycle. Ash and sulfur are separated from the coal at the proper time—before burning. Upon combustion of low-BTU gas no smoke is released, and the emission of acid-forming sulfur oxides should be negligible, as the gas contains less than one-tenth parts per million of sulfur. Gas turbines fired with low-BTU gas will also generate much less nitrogen oxides than coal furnances. The amount of energy wasted in generating electricity could be halved, as the thermal efficiency of a combined-cycle arrangement can reach 60%. Finally, this type of power plant can be situated in the middle of a new community (since it does not pollute the air) where it can pipe out its "waste" heat to homes for space heating and cooling.

Refined coal is also desirable as a clean fuel since it may be burned in existing power plants. It does not achieve the same low levels of sulfur content as gasified coal, but a reduction to 0.2% sulfur is remarkable, nonetheless. The major outstanding question about refined coal is whether it can be burned with acceptable emissions of nitrogen oxides.

Synthetic pipeline gas will probably be flowing through the pipelines before the year 2000, because our reserves of natural gas are not likely to be sufficient for meeting projected demands by then. But the nation should be cautious about switching on a large-scale from natural to synthetic gas. Environmental degradation will occur at two places— the gasification plant and the coal mine. The plant will discharge waste heat, sulfur ·dioxide, and extra carbon dioxide. Complete replacement of natural gas and coal (as ultimate fuels) with synthetic gas would require that the extraction of coal increase $3\frac{1}{2}$ times over present levels, even if there is no growth in demand. It raises the prospects of intensified strip mining of coal on a scale that this country has never imagined. Synthetic fuel is coming soon, and it should be welcomed—provided we demand environmental safeguards.

REFERENCES

Akhtar, Sayeed, 1971. *Low-sulfur fuel oil from coal.* Bureau of Mines Technical Progress Report 35, July.

Coal News, No. 4071, Aug. 20, 1971.

Forney, A. J., et al., 1970. *A process to make high-BTU gas from coal.* U.S. Bureau of Mines Technical Progress Report 24, Apr.

Lee, B. S., 1971. The development of the HYGAS Process for converting coal to synthetic pipeline gas. Institute of Gas Technology, Oct. 3.

Lyzenga, David R., 1971. Patterns in energy consumption. Space Research Laboratories, Univ. of Michigan.

Mills, G. Alex, 1970. Progress in gasification—U.S. Bureau of Mines. Presented to Third Synthetic Pipeline Gas Symposium, Chicago, Nov. 17.

Stewart, E. G., 1958. Town gas—Its manufacture and distribution. Science Museum, London.

United Aircraft Research Laboratories, 1970. *Technological and economic feasibility of advanced power cycles and methods of producing nonpolluting fuel for utility power stations.* Report No. 970855-13, Dec.

U.S. Office of Coal Research, 1968. Project H—Coal report on process development. R&D Report No. 26.

————, 1970a. Economic evaluation of a process to produce ashless, low-sulfur fuel from coal. R&D Report No. 53, Interim Report No. 1.

————, 1970b. Phase II, bench scale research on CSG Process. R&D Report No. 16, Interim Report No. 3, Books 1, 2, and 3, Jan.

————, 1970c. Final report of the Advisory Committee on Project Gasoline. R&D Report No. 62.

————, 1971a. Desulfurization of COED Char—Part III. R&D Report No. 56, Interim Report No. 2, Jan.

————, 1971b. Engineering study and technical evaluation of the Two Stage Super Pressure Gasification Process. R&D Report No. 60, Feb.

————, 1971c. *Annual Report.*

————, 1972. *Annual Report.*

Wen, C. Y., et al., 1972. Comparison of alternate coal gasification processes for pipeline gas production. Paper presented at the New York meeting of the American Institute of Chemical Engineers, Nov.

TERRI AARONSON

The Black Box:
The Fuel Cell[1]

A clean, safe means of generating electricity may result from the development of the fuel cell, a kind of "black box" which has virtually no moving parts, makes no noise, is free of vibration, and generates only innocuous products and electricity. The fuel cell is actually an electrochemical device; that is, it converts the chemical energy of various fuels directly into electrical energy. Ordinarily, heat produced by burning fuel or by nuclear reaction is used to produce steam to turn large generating units that produce electricity. This process of using chemical energy to produce heat energy to produce mechanical energy to produce electrical energy is inherently inefficient. The newest conventional generating devices operate at about 40% efficiency (Angrist, 1971). Fuel cells theoretically approach 100% efficiency in converting chemical energy to electrical energy (Bockris and Srinivasan, 1969). Because of their greater efficiency, fuel cells require less fuel to generate an equal amount of electricity than do conventional or nuclear power plants. Another advantage of fuel cells is that they remain efficient over a wide range of power levels and can be built in large or small units. Fuel cells are being developed to relieve peak power loads for conventional power stations, for vehicular use, and for potential use as large (1000-megawatt) central power stations. A major project (discussed below) is now under way to test the feasibility of using small fuel-cell units to provide the entire energy needs of an individual home, apartment complex, or industrial area.

Electric power generation contributes significantly to the degradation of air quality throughout the nation. In 1966 it was estimated

[1]Reprinted, with permission, from *Environment*, Vol. 13, No. 10. December 1971.

(U.S.P.H.S., 1966) that electric power plants contributed a total of twenty million tons of sulfur oxides, nitrogen oxides, particulate matter, carbon monoxide, and hydro-carbons to the atmosphere. Nuclear power plants emit low levels of radiation to the atmosphere, generate larger amounts of radioactive wastes, and contribute to thermal pollution by the need for vast quantities of cooling water that are discharged to the environment at high temperatures. The use of fuel cells for power generation would avoid producing the above-mentioned pollutants. And, by using fuel cells in automobiles the major source of air pollution in urban areas—the internal combustion engine—would be eliminated.

The fuel cell was discovered by Sir William Grove in 1839 (Liebhafsky and Cairns, 1968), but the devices have become practical only within the past decade. Grove's fuel cell consisted of separate tubes of hydrogen and oxygen in a dilute solution of sulphuric acid. Grove placed strips of platinum foil into the tubes of gas and recorded that "a shock was given which could be felt by five persons joining hands, and which when taken by a single person was painful" (Liebhafsky and Cairns, 1968). After Grove's initial experiments, though, little attention was paid to fuel cells because of the costliness at that time of producing hydrogen and oxygen, in addition to the expense of the platinum. Fifty years after Grove's first fuel cell, which was intended solely for limited scientific purposes, Ludwig Mond and Carl Langer recognized the fuel cell's potentiality as a new means of generating electricity. Although Mond and Langer developed a device superior to Grove's first fuel cell, it was overshadowed by the steam-powered dynamo. Work on electrochemical cells once again proceeded slowly, with researchers looking for fuels other than hydrogen.

Modern fuel-cell technology was pioneered by Francis T. Bacon in the 1930s. In 1959 Bacon announced that he and J. C. Frost of Cambridge University had developed the first practical fuel cell. It could be used to power a forklift truck, a circular saw, and a welding machine. Just two months after Bacon's announcement, H. K. Ihrig of Allis-Chalmers Mfg. Company demonstrated a fuel-cell tractor that he had developed.

The biggest boost to fuel-cell development, though, came when it was realized that, because of its high power output and low weight, the fuel cell was well suited to space applications. Government-sponsored research on fuel cells soared, reaching a high of $15.9 million in 1963 (Kordesch, 1970). However, government interest has waned, and it seems that fuel-cell research has been shelved. In 1969 only $2.2 million in government funds was spent on fuel-cell research, climbing to slightly more than $3 million for 1970 (Kordesch, 1971d).

This seeming lack of interest on the part of the government is somewhat ironic in light of professed concern with pollution and the reference, made by William Ruckelshaus, administrator of the federal Environmental Protection Agency, to the "enormous unrealized potential" of fuel cells (Ruckelshaus, 1971).

VERSATILITY OF FUEL CELLS

A single fuel cell technically is only a single set of two electrodes, joined by an electrolyte (which conducts current between the two electrodes) and an external circuit. Each set of electrodes is comprised of a fuel (such as hydrogen or natural gas) electrode and an oxygen (for air) electrode; thus, a hydrogen/oxygen fuel cell is one that has a hydrogen electrode and an oxygen electrode (see later discussion of fuel-cell components). Only a small amount of current can be drawn from a single fuel cell. A primary attraction of fuel cells is that they can be joined together in series or parallel to attain the amount of power and voltage desired for a particular task. There is no loss of efficiency in joining a great many fuel cells together, nor is there a significant reduction in efficiency by joining only a few cells together. Since fuel cells can be made very thin, they can be stacked occupying little space. When connected together, they technically are called a fuel-cell battery, or sometimes a fuel-cell power plan. (Fuel cells described herein are actually fuel-cell batteries.)

Because the power output of fuel cells can be adjusted by adding or subtracting units, fuel cells are extremely versatile. They have been envisioned for every size power plant imaginable. Domestic units may someday provide all the electricity an individual home might need. Or, a moderate-sized unit might provide power to a single neighborhood. Or, a slightly larger unit might power an entire community in a remote area. Or, if it is more convenient, very large fuel-cell power plants could be assembled to provide large blocks of power to whole areas, much as the present conventional power plants do.

Fuel cells can provide electricity for a host of uses other than power plants. The Army, for instance, is interested in fuel-cell applications for small generating units to be used in the field. Fuel cells are attractive tactically because they make no noise and because they can be operated at low temperatures that are undetectable by infrared sensors. Fuel cells have proved to be efficient in space applications, and someday biochemical fuel cells may offer a simple solution to the problem of human waste disposal from spacecraft, by transforming the wastes into electrical energy. Another intriguing possibility for biochemical fuel cells is a pacemaker for persons with deficient hearts; such a pacemaker would run off of substances found within the body. It has been suggested that blood can provide the fuel and necessary oxidant, in addition to acting as an electrolyte, for a totally implanted pacemaker (Angrist, 1971).

Fuel cells have already been installed in a variety of vehicles, from forklift trucks to tractors, and may someday be used to power non-polluting automobiles, buses, trains, or even submarines. In the mid-1960s the General Motors Corporation designed and built a van that was powered by hydrogen/oxygen fuel cells. The van was estimated to compare favorably with a conventional van in terms of road tests (Marks et al., 1967), but the "electrovan" was not without disadvantages. Most notably, the electrovan was heavier than the conventional one, and the fuel-cell materials were expensive. These disadvantages may be overcome by future development, and GM is continuing to work with fuel cells. A more difficult problem, though, is the hazard of storing high-energy fuels under pressure. Safety requirements similar to those stipulated for commercial gasoline trucks might be necessary for a hydrogen/oxygen fuel-cell vehicle.

A more recent fuel-cell-powered vehicle that has bypassed some of the difficulties of the electorvan has been designed and built by K. V. Kordesch of the Consumer Products Division of the Union Carbide Corporation. This vehicle is actually powered by a hybrid system composed of a fuel cell and a conventional battery. Currently the car contains a hydrogen/air fuel-cell power plant and several lead-acid batteries for additional short bursts of power (Kordesch, 1971a). A potential for overcoming the explosion hazard of the fuel rests in the possibility of operating the car with an ammonia/air fuel-cell power plant, again supplemented by batteries (Kordesch, 1971c). Liquid ammonia does not need to be stored under high pressure, and thus is less subject to explosion. However, escaping liquid ammonia, like hydrogen, is dangerous in a closed space, the latter due to dangers of asphyxiation.

The fuel cell/battery hybrid vehicle is an Austin sedan that was originally converted from an internal combustion engine to an all lead-acid battery power source. With the present hybrid system, the car presents a possible alternative to polluting automobiles that are now driven in urban areas. The hybrid car can run for 200 miles without refueling, and has a top speed of 50 miles per hour, thus making it a feasible urban vehicle, but not acceptable for long-distance driving until popularity of the vehicle could eliminate difficulties in refueling. Dr. Kordesch, who often drives the car for personal use, has logged more than 2500 miles on it since its conversion to fuel cells in 1969. The features of the car (Kordesch, 1971b) suggest that such a power-plant design is quite practical, although widespread acceptance is unlikely in the near future considering that the proven internal combustion engine will probably be cleaned up sufficiently to meet projected air pollution abatement standards. However, as standards become more stringent, and more controls are placed on the combustion engine, the cost and efficiency of a fuel-cell-powered vehicle may appear more attractive.

The fuel cell/battery car takes one to two minutes initial start-up time for the fuel cell to attain full power. While the fuel cell is "warming up," the batteries can be used to drive the car. The batteries can be recharged by the fuel cell at a stop—a traffic light, for instance. Recharging the hydrogen tanks carried atop the car is little problem, with an average time of three minutes to refill the tanks. And car maintenance should be minimal. The fuel cells themselves present little problem, and because they are completely shut down during long off periods (the car is designed so that the electrolyte drains from the fuel cells during lengthy shutdown) and little power is lost during idling, the fuel cells should remain capable of producing sufficient power for quite some time. If fuel cells for automobiles were mass produced, the cost of the cells might be acceptable, and the cost of running the car would be small, an estimated one-half cent per mile.

TWO POWER-PRODUCTION PROJECTS

The electric utility industry seems to be watching the development of fuel cells, but is not involved in any large-scale projects to develop fuel-cell technology. However, development of fuel cells for commercial power production is not at a standstill. Two projects are testing the feasibility of fuel cells for generation of electric power on a mass scale. The first project is sponsored by the Office of Coal Research (OCR), an agency of the federal Department of the Interior. The second, and larger, project is sponsored by a coalition of gas utilities and the Institute for Gas Technology.

OCR has spent approximately $3.75 million during the past seven years on developing a fuel cell that will be run on coal. Although there was no active work on a coal-reacting fuel cell in the past year, it is likely that Westinghouse Electric Corporation (OCR's prime contractor) will soon begin construction of a very small-scale (100-kilowatt) power plant that will test the practicality of a complete coal-reacting power plant. The coal-reacting power plant is actually an "indirect" fuel cell: the coal is first gasified (producing hydrogen) in a high-temperature gas generator; the hydrogen is then reacted with air to produce water and electricity. Sulfur is scrubbed from the gas before it enters the fuel cell, and nitrogen oxides are not produced. By-products of this type of fuel cell are carbon dioxide and water. The fuel cell itself utilizes a solid electrolyte, and because it is operated at high temperatures, it is not expected to need expensive "noble" metal catalysts. (Noble metals are those such as silver, platinum, and gold that are good conductors of electricity.) OCR has estimated that a complete fuel-cell power system utilizing coal will be 60% efficient, and that nearly one and one-half times more electrical energy can be obtained from each ton of coal used in a fuel cell than in a conventional plant (OCR, 1971). Thus, fuel cells would contribute greatly to conserving our coal resources. Other advantages of the coal-reacting fuel cell are that it does not require cooling water (high temperatures generated in the fuel cell can be used effectively in the coal-gas generator), and thus sites for fuel cells should not be dependent upon the availability of large bodies of water. Fuel cells have no moving parts, and thus should prove reliable in operation. And, they have a high power density per cubic foot; it is estimated that a sufficient number of fuel cells to generate one megawatt of power would fit into an area seven feet in diameter and six feet high (OCR, 1971).

If OCR is successful in demonstrating that coal-reacting fuel cells are a practical means of producing energy, it is possible that such cells will be built. Economics will dictate that these power stations be large and centrally located and that the electricity produced be transmitted over the present grid of high-tension wires. Or, it is possible that by the time central fuel-cell power plants are built, new, more efficient transmission lines, as are now envisioned, will have come into existence. However, no matter how the energy is transmitted, some energy will be lost in the process, thus cutting down on the efficiency of the total system of power generation and delivery.

The most obvious way of avoiding loss of energy by transmission is to produce power near the consumer. That is, in fact, the goal of the project that is being pursued by the Team to Advance Research for Gas Energy Transformation, Inc., known as TARGET, a coalition of 32 member companies, most of which are gas utilities. Together with the Pratt & Whitney Aircraft Division, TARGET has developed a fuel-cell power system which is designed to be used for localized power production (Connecticut Natural Gas Corp.). Now in its fourth year, the TARGET program is in a stage of on-site testing. Plans call for fuel-cell power plants to be installed in 37 locations throughout the country. The fuel cells will provide power for homes, apartments, stores, restaurants, and office and industrial buildings.

Each fuel-cell system has the capability of generating 12.5 kilowatts of power; that should be more than sufficient to meet the peak demands of an individual dwelling unit. For the industrial installations, it may be necessary to arrange numerous fuel-cell systems in series and parallel to provide adequate power. There will be no loss of power or efficiency with multiple cells.

TARGET's fuel-cell system runs on natural gas. It could, of course, be run on a synthetic gas. The system is again an indirect fuel cell, using a steam reformer to decompose the natural gas (methane) to hydrogen and carbon dioxide. The gases are then fed into the fuel cell itself, which acts as a hydrogen/air cell. Water formed as a product of the fuel-cell reaction is circulated back to the steam reformer. Electricity produced by the system is direct current power, as is the power produced by all fuel cells. Currently the TARGET fuel cell is attached to an inverter which transforms the direct current into alternating current. Losses of efficiency occur in both the reformer and in the inverter; the remaining over-all efficiency of the TARGET system is about 45%—equal to or better than other generating systems (not including losses during transmission). However, many applications of electric power are more favorably met with direct current.

The TARGET fuel-cell system, like other fuel cells, is very clean, producing carbon dioxide, water, and some heat as products. It is quite compact, about the size of a furnace, and commercial models are expected to be reduced to about the size of a gas air conditioner. Because there are very few moving parts (pumps and fans) in the fuel cell, it should produce practically no noise, and should be quite reliable. The efficiency of the TARGET fuel cell is expected to permit the system to provide a given amount of power to a site using one-third less fuel than any other power system. The cost of the system should be reasonable because, although the initial installation of a fuel cell may be relatively expensive, the savings on fuel and the low cost of gas transmission (as compared to electricity transmission) should more than compensate installation costs. Transmission and distribution of energy in the form of gas in most parts of the country cost only one-fifth as much as the transmission of an equal amount of energy in the form of electricity. However, the economics of the entire TARGET system have not yet been worked out in detail; projected costs will not be estimated until completion of the installation program, probably in late 1972. It has been suggested, though, that "successful large-scale use of fuel cells will depend more on resolving economic problems than technical ones" (Connecticut Natural Gas Corp.). Technical problems seem to have been minimal in the first installation of the TARGET program. A fuel cell was installed in a model home in Connecticut in July and ran through September. Full evaluation of the first practical test is not complete, however.

COMPONENTS OF A FUEL CELL

The basic fuel cell consists of two electrodes—the anode, or fuel, electrode; and the cathode, or oxygen, electrode (air may be substituted for oxygen). Both the fuel and oxygen are continuously fed to the electrodes, which need some sort of catalyst for accelerating the chemical reactions. Electrode catalysts can be made of many materials; however, most fuel cells work best with noble metals, usually platinum. Unfortunately, platinum and the other noble metals are relatively scarce resources, making them quite expensive. Thus, the need for noble metals in fuel cells is one of the most difficult problems to overcome before fuel cells can be practical for widespread use. The one other main component of a fuel cell is the electrolyte, which conducts current between the two electrodes. Electrolytes may be many substances, either acid or alkaline, liquid or solid.

The modern fuel cell is similar to Grove's original cell. Fuel is fed to the anode (which, in the case of fuel cells, is negative, unlike the ordinary electrolysis battery) where the fuel is oxidized in the technical sense, i.e. gives up electrons, thereby forming ions of the fuel substance. The electrons travel along an external metallic path, or circuit, to the cathode. The energy of the fuel cell is tapped by placing some sort of electric load in the circuit. At the cathode either oxygen or air normally is fed to the fuel cell, although other reducing agents may be used. The oxygen combines with the ions of the fuel that have formed at the anode and with electrons to form a new compound that can be drawn off from the cell. For an illustration of a simple hydrogen/oxygen fuel cell, see Figure 1.

Many fuels other than hydrogen can be used in simple fuel cells. Some combinations other than hydrogen/oxygen that are being considered for fuel cells are: hydrogen/air; carbon or carbon monoxide/air; natural gas (methane)/air; hydrazine/oxygen or air; sodium amalgam/oxygen; zinc/oxygen; and aluminum/oxygen. Not all of these combinations have yet been proven to be practical. Further considerations are the choice of materials for the electrode catalysts and the electrolyte. None of the combinations mentioned above are expected to produce uncontrollable pollutants of any sort. For instance, when air is used in place of oxygen, the nitrogen in air does not form harmful nitrogen oxides, but rather is emitted simply as nitrogen gas.

The most glamorous use of fuel cells to date has been in space applications. In the Gemini and Apollo spacecraft, hydrogen/oxygen fuel cells were used. Quite naturally, then, this type of fuel cell is currently the most technically advanced. However, it is unlikely that the fuel-cell design used for either the Gemini or Apollo programs

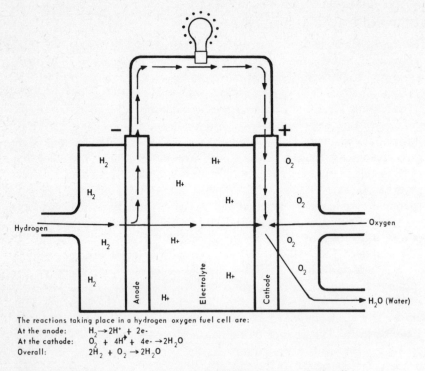

The reactions taking place in a hydrogen oxygen fuel cell are:
At the anode: $H_2 \rightarrow 2H^+ + 2e^-$
At the cathode: $O_2 + 4H^+ + 4e^- \rightarrow 2H_2O$
Overall: $2H_2 + O_2 \rightarrow 2H_2O$

Figure 1   Electricity from a simple hydrogen/oxygen fuel cell

will ever be used on a mass scale. The Apollo fuel cell has a relatively short lifetime, and the Gemini fuel cell needs expensive platinum catalysts for the electrodes. Recent developments, though, have reduced to a very low level the amount of noble metal catalysts necessary for the operation of a fuel cell, and it has now been shown that non-noble metal catalysts, such as nickel compounds, borides, carbides, and organic catalysts, can work efficiently in some fuel cells (Kordesch, 1971c).

KINDS OF FUEL CELLS

Rather than using pure hydrogen (which has limited availability) in a fuel cell, it may be advantageous to use some hydrocarbon compound, such as natural gas, which, for example, is used in the TARGET program. Natural gas is already widely available through a nationwide network. Although resources of natural gas are declining, it is relatively inexpensive, and a synthetic natural gas can be produced from a variety of materials, including coal and sewage. There are two kinds of fuel cells that can use natural gas as a fuel—the direct cell and the indirect cell.

The direct fuel cell is one in which the hydrocarbon fuel is fed to the anode, in either a gaseous or liquid state, and then the fuel cell operates as described above, except that carbon dioxide is formed as a reaction product and must be drawn from the cell so that it does not clog the workings of the electrodes. Direct hydrocarbon fuel cells are still in the early developmental stages but have already been shown to be more efficient than conventional engines using the same fuels. See Bockris and Srinivasan (1969) for review of current work on both direct and indirect fuel cells. However, until electrode catalysts other than the noble metals can be found that will operate in direct hydrocarbon fuel cells, the direct cells will probably be too expensive for common acceptance.

Indirect fuel cells involve a process by which the hydrocarbon undergoes a steam-reforming process outside of the cell to obtain pure hydrogen which is then fed into the fuel cell. (More energy is produced from the fuel cell than is required to decompose the hydrocarbon.) Once the hydrogen is produced, the fuel cell functions as any ordinary hydrogen/air cell. The main advantage of the indirect fuel cell is that it does not require costly platinum electrocatalysts, but can operate with less expensive materials. The primary disadvantage of the indirect cell is that the reforming process requires energy which cuts down on the efficiency of the cell.

Two fuel cells which also seem very promising use a nitrogen-hydrogen compound, either hydrazine or ammonia, as fuel. Hydrazine is an attractive fuel because it is highly reactive and it does not require expensive electrocatalysts. Unfortunately, hydrazine itself is very expensive and toxic. There are indications, however, that large-scale production methods of fabricating hydrazine may bring down the cost. Even if that is not accomplished, the high energy output of hydrazine is likely to allow the fuel to be used in specialized fuel-cell applications (Liebhafsky and Cairns, 1968). Unlike hydrazine, ammonia is quite inexpensive. The performance of ammonia in fuel cells, is however, markedly inferior to hydrogen/air cells (Bockris and Srinivasan, 1969). Therefore, it seems logical that if ammonia is to be used as a fuel, it should be an indirect fuel cell, which would first reform the ammonia to hydrogen and nitrogen gas.

Some fuel cells have been designed so that the products of the cell can be decomposed into the initial fuel and oxident; for example, water may be disassociated into hydrogen and oxygen. These fuel cells are called regenerative. In order to regenerate the fuel, energy must be added to the system. This can be accomplished by thermal, electrical, chemical, radioactive, or photochemical methods. Most research on regenerative fuel cells has been done on electrically regenerative cells using lithium/chlorine or hydrogen/bromine as fuel combinations. The substances are fairly expensive and thus are practical only if they are reused. Electrically regenerative fuel cells are actually a kind of rechargeable battery, in that they must be fed electricity to continue working; as such, if they were to come into large-scale use (they have been suggested for use in replacing the present highly polluting automotive engine) (Bockris and Srinivasan, 1969), they would necessitate greatly enlarged capacities for producing electric power. Unless that power were produced by fuel cells or some other clean method, electrically regenerative fuel cells could intensify, rather than alleviate, pollution by central power stations.

Regenerative fuel cells other than those powered by electricity have not yet been carefully studied, but other means of regeneration present some interesting possibilities. Radiochemical regenerative fuel cells would work by radiolysis, that is, the product of the fuel cell would be dissociated by irradiation. J. O'M. Bockris, of the University of Pennsylvania, has suggested that fuel cells could use radioactive nuclear wastes as a means of regeneration. Because the regenerative fuel cell is a kind of closed system, little or no radiation should escape. However, a radiochemical regenerative device has yet to be proven powerful enough to be practical, much less safe.

The only other type of regeneration that seems plausible currently is photochemical regeneration. Such a system would utilize solar energy to transform the fuel-cell product to its reactants. Photochemical regeneration is most likely to find uses in space applications, where exposure to solar energy is assured.

One of the most fascinating fuel-cell concepts is that of a biochemical cell. In the biochemical fuel cell the reactions at one or both of the electrodes are promoted or catalyzed by biological processes (Liebhafsky and Cairns, 1968). Although the biochemical fuel cell sounds at first outlandish, it can be an efficient reliable process. It is the process, in fact, that powers the human body. In the body, enzymes catalyze reactions wherein food (fuel) is oxidized in cells to produce energy, some of which is electrical (Mitchell, 1963). Biochemical fuel cells may be either of the direct or indirect types. The indirect biochemical fuel cell operates on the waste products of a living organism. Fuel (such as formic acid) is fed to bacteria (such as Escherichia coli) that then produce a waste product (such as hydrogen) which can be used directly as the anode fuel in a fuel cell (Williams, 1966). Conversely, algae can produce oxygen for the cathode. Or, an indirect fuel cell can be one in which an enzyme acts on a substance that is then fed into the cell.

Direct biochemical fuel cells are ones in which all the reactions take place within the fuel cell. For example, a direct biochemical fuel cell might use glucose as the fuel electrode, with the enzyme glucose xoidase as the electrode catalyst. Or, a direct biochemical fuel cell might actually have living organisms within the fuel cell to provide the enzymes required by the biochemical-electrochemical process. One of the most interesting biochemical fuel cells is one designed to run on urea (in urine) and air, catalyzed by the enzyme urease. It is possible that such fuel cells, or others designed to run on organic wastes, might be useful in remote areas of underdeveloped nations. However, it has yet to be proven that any feasible biochemical fuel cell can produce a significant amount of power economically.

ECONOMICS

Evaluating the cost of fuel cells for various applications is largely a matter of conjecture at this stage of development. However, it is possible to project expected costs of fuels and to foresee reductions in the costs of fuel-cell manufacturing as the technology is advanced. The most comprehensive review of fuel-cell economics to date suggests that fuel cells are most likely to find application in the large-scale production of power (Verstraete, 1968). Fuel cells may also be used economically in small, specialized vehicles such as forklift and delivery trucks and in other specialized areas such as railway engines and naval propulsion. General automobile applications of fuel cells are not at present or in the near future economically attractive. In comparison to conventional engines, fuel cells in the power range considered for automobiles are most economical when used intensively. Thus, it would be more economical to use fuel cells for taxicabs than for general-use automobiles; however, even taxi applications do not yet compare favorably with the internal combustion engine. According to the review mentioned above, for fuel cells to become appealing for automotive use, the cost of fuel would have to be drastically lowered or the cost of manufacturing the fuel cells would have to be reduced beyond present estimates. Dr. Kordesch (1971c), however, expects that if fuel cells were mass produced for automobile use, they would become economically competitive with the internal combustion engine. Catch 22, though, is that Kordesch does not envision fuel cells being mass produced until their cost is reduced. The one aspect that may confuse all these projections is that future air-quality standards may require pollution controls on the internal combustion engine which may lower the efficiency and increase the cost sufficiently to make fuel cells competitive, or even preferable, economically. Reduction in the supply of fossil fuels may further enhance the economics of fuel-cell use.

Table 1 Comparison of Fuel Cell Costs with Other Means of Electric Power Production (in Cents per Kilowatt-Hour)

Types of Costs	Fuel Cell Plants		Fossil Fuel Plants		Nuclear Water		Fast Breeder	High Temperature Gas Turbine	Magnetohydrodynamic Power Plant
	1980[a]	1990[b]	1980	1990	1980	1990	1990	1990	1990
Annual Capital	0.67¢	0.34¢	0.54¢	0.67¢	0.84¢	1.05¢	1.05¢	0.62¢	0.62¢
Fuel	0.36¢	0.36¢	0.57¢	0.71¢	0.23¢	0.29¢	0.10¢	0.48¢	0.48¢
Maintenance and Operation; Insurance	0.04¢	0.04¢	0.03¢	0.04¢	0.06¢	0.07¢	0.07¢	0.04¢	0.04¢
Thermal Effects	0.02¢	0.02¢	0.03¢	0.04¢	0.04¢	0.06¢	0.04¢	0.03¢	0.03¢
Total	1.09¢	0.76¢	1.17¢	1.46¢	1.17¢	1.47¢	1.26¢	1.17¢	1.17¢

[a]These figures hypothetically consider that the initial capital cost equals that of fossil fuel plants in 1990.
[b]These figures consider that the high efficiency and ease of manufacturing of fuel-cell plants brings their initial cost to one-half that of fossil fuel plants.

Source: Baron, S., "Options in Power Generation and Transmission" presented at Brookhaven National Laboratory Conference, "Energy, Environment, and Planning, The Long Island Sound Region," October 5–7, 1971.

Large fuel-cell power plants, on the other hand are predicted to be economically attractive by their proponents, even if future air pollution regulations are not considered. Depending on the size and type of the fuel-cell power system, fuel-cell electric power may cost as much as 0.627 cent per kilowatt-hour less than power from future conventional power plants (Berger, 1968). Another analysis of fuel-cell power costs, considered only for very large central generating stations, estimates that, depending on the cost of initial investment for fuel cells, fuel-cell-generated power may cost as much as 0.71 cent per kilowatt-hour less than some other means of producing large blocks of electricity (Table 1), and that even with high estimates of initial investment, fuel-cell-generated power will be less costly than light- or boiling-water reactors, fast breeder reactors, high-temperature gas turbines, magnetohydrodynamics, or fossil-fuel power plants (Baron, 1971). Estimates made for the Office of Coal Research indicate that a coal-fired fuel-cell plant of large size (one million kilowatts—or 1000 megawatts) will provide energy for only 0.299 cent per kilowatt-hour (OCR, 1966)—a cost far below the one indicated in Table 1. This discrepancy emphasizes the hazards in projecting cost for new technologies when there are so many variables.

Cost of fuel cells will undoubtedly decrease, though, as the technology advances. New, lower cost electrodes will be developed, and costs for electrolytes, particularly solid ones, will decrease as the science is perfected. There are certain inherent advantages, too, that will help fuel cells become economically competitive with other means of power production. The efficiency of fuel cells permits them to use less fuel to produce a given amount of power than do other methods. As fuel prices increase with time, as they are projected to do, the fuel cell's efficiency will be emphasized, and expected increases in the practical efficiency of fuel cells will enhance this aspect. But the most significant advantage of fuel cells is their inherent cleanliness. The products of fuel cells are usually water, carbon dioxide, and nitrogen. Of these products, the only one that might have a detrimental effect on the environment is carbon dioxide, which has been suggested by some scientists to cause an atmospheric "greenhouse" effect, which may produce weather changes. However, the amount of carbon dioxide released from fuels, be they used in conventional power generators or in fuel cells, is constant. Therefore, fuel cells would release no more carbon dioxide into the atmosphere than do conventional generators; in fact, fuel cells would probably release less because they need less fuel to produce an equal amount of energy.

Fuel cells seem to hold the answer to the question of how to produce energy economically, at least for some major uses. Although the technology of fuel cells has improved vastly in the last decade, more developmental work must be done before these devices are commercially practical. It is somewhat ironic that, in these times of increasing environmental concern, little governmental funding for fuel-cell development is forthcoming.

REFERENCES

Angrist, Stanley W., 1971. *Direct energy conversion.* 2nd ed., p. 20. Boston: Allyn and Bacon, Inc.

Baron, S., 1971. Options in power generation and transmission. Presented at Brookhaven National Laboratory Conference, Energy, Environment and Planning, The Long Island Sound Region, Oct. 5–7.

Berger, C., ed., 1968. *Handbook of fuel technology.* Englewood, N.J.: Prentice-Hall.

Bockris, J. O'M., and S. Srinivasan, 1969. *Fuel cells: Their electrochemistry.* St. Louis: McGraw-Hill Book Company, p. 2.

Connecticut Natural Gas Corporation. Target for tomorrow. Hartford, Conn.

Kordesch, K. V., 1970. Outlook for alkaline fuel cell batteries. Battelle Institute Seminar. *From Electrocatalysis to Fuel Cells,* Seattle, Dec. 9–11.

———, 1971a. Hydrogen-air/lead battery hybrid system for vehicle propulsion. *J. of the Electrochem. Soc.* 118, no. 5: 812–817.

———, 1971b. City car with H_2-air fuel cell/lead battery (one year operating experiences). Society of Automotive Engineers, Inc., 1971 Intersociety Energy Conversion Engineering Conference Proceedings, p. 38.

———, 1971c, personal communication, Oct. 12.

———, 1971d, personal communication, Oct. 29.

Liebhafsky, H. A., and E. J. Cairns, 1968. Fuel cells and fuel batteries, A guide to their research and development. New York: John Wiley & Sons, Inc., pp. 18, 19.

Marks, C., et al., 1967. Electrovan—A fuel cell powered vehicle. Society of Automotive Engineers, Detroit, Automotive Engineering Congress and Exposition, January 9–13.

Mitchell, Will, Jr., 1963. *Fuel cells.* New York: Academic Press, p. 1.

Ruckelshaus, W. D., 1971. Energy and environment. Address to World Energy Conference, Washington, D.C., Sept. 24.

U.S. Office of Coal Research, 1966. News release. Department of the Interior, Washington, D.C., Dec. 12.

———, 1971. *Annual Report,* p. 48.

U.S.P.H.S., 1966. Sources of Air Pollution and Their Control, Public Health Service Publication No. 1548, Washington, D.C.

Verstraete, J., et al., 1968. Fuel cell economics and commercial applications. In Berger, 1968.

Williams, K. R., ed., 1966. *An introduction to fuel cells.* New York: Elsevier Publishing Company, p. 254.

RICHARD C. BAILIE

Wasted Solids as an Energy Resource

The U.S., as a nation, has made little effort to utilize as an energy or mineral resource, the solid waste produced from our affluent society. In the past it has been easier to utilize our irreplaceable natural resources while discarding a valuable portion of these resources rather than to put forth the effort necessary to utilize these waste materials.

Only recently has there been a general realization that our nation, although rich in resources, can no longer continue to think only of today's needs. We must look at the future and consider the effect our policy of unconstrained utilization of virgin resources will have on future generations. With this new realization, newly developed technology, and the need to protect our environment, it becomes important that we consider the impact of solid waste as a potential natural resource.

It is the purpose of this paper to assess the potential of solid waste as a readily available energy resource. The paper discusses the generation of solid waste along with its potential as an energy source and reviews some of the processes that are being developed which could result in effective recovery of the fuel value of solid waste. Some of the unique problems associated with solid waste that may retard any progress toward its effective utilization as a fuel are also discussed.

ENERGY POTENTIAL OF SOLID WASTE

Figure 1 shows present estimates of gas, oil, and coal utilization to furnish the nation's energy needs and the total energies available from three major classifications of solid waste (municipal, agricultural, and animal). It can be seen that the energy available from these three solid wastes is substantial and represents a potential fuel resource.

117

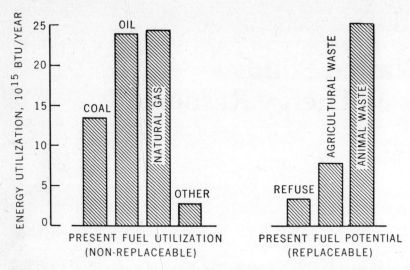

Figure 1 Fuel utilization and potential in the U.S. by source, 1970

It has been pointed out by several persons (Clark, 1971; Mills and Johnson, 1971; Glaser, 1968) that ultimately our energy consumption must come directly from the sun. Many unique proposals have been described to utilize the sun's energy.

A major portion of solid waste is cellulose and is, therefore, derived from "today's" photosynthesis. This is in contrast to our fossil fuel resources which represent "yesterday's" photosynthesis. Thus, if we are able to utilize our solid waste as an energy resource, we are, in effect, making use of the sun's energy which brings us closer, energy-wise, to living within our means. Since solid waste is generated from the incomplete utilization of our natural resources, the energy available from solid waste is then, in effect, a "free" energy.

Although the total energy potential of our nation's solid waste is substantial, it is not feasible to consider it all useable as a fuel. Solid waste is not found concentrated in areas, as is the case with other fuels, but is dispersed throughout the nation. For this reason, it is not reasonable to consider utilization of a major fraction of the animal and agricultural wastes. However, the picture is changing. The direction is toward larger and larger concentrations of these animal and agricultural wastes in smaller and smaller areas. The solid waste picture parallels that of the nation's changing population where the rural population is decreasing while the urban population is growing. For example, there are presently over 3500 cattle lots having a cattle population of over 1000 head. Feed lots of 50,000 head are equivalent to 500,000 persons in solid waste production and are not uncommon. The trend toward high population density centers where solid waste is collected and transported to some common location makes municipal waste an important candidate as a fuel resource.

By only considering the fraction of the nation's energy requirements that can be supplied by solid waste, it can be seen that the utilization of municipal waste as an energy resource is limited, but this is not sufficient basis for assessing solid waste's potential. By utilizing the energy of this waste material, we will have made a giant step forward toward the solution of our nation's solid waste pollution problem.

It is in the metropolitan areas of the east and west coasts where fuel costs are highest and SO_2 emissions must be curtailed. These regions are simultaneously approaching a solid waste disposal crisis along with a shortage of low sulfur fuel. In effect, these areas are being buried under the very thing they so desperately need, a low sulfur fuel.

The amount of municipal solid waste available as a fuel is growing. The factors that affect this growth are:

1. growth of large cities where waste is collected
2. growth of the per capita generation of waste
3. the increase of the solid waste heating value

These trends are shown in Figure 2. Two indicators that may be used to indicate the affluence of a modern society are the per capita energy consumption and the per capita waste generation. Our nation is leading the rest of the world in both of these categories (as a nation with 7% of the world's population, we utilize over half of the world's resources). We must begin to reduce these two indicators by better conservation and utilization of our resources, including our solid waste.

A typical analysis of municipal solid waste is presented in Table 1. These values are average values and must be interpreted with caution, since these values vary widely from region to region and season to season. The heating value of waste is compared to other fuels in Figure 3, and it can be seen that the heating value for solid waste is in the range of 3000–6000 BTU per lb. The largest constituent of solid waste is cellulosic in nature with a heating value of 8000 BTU per lb. The high inert solid (glass, metal, etc.) content and moisture content of solid waste drastically lower its heating value.

A break down of the chemical composition of solid waste is presented in Table 2, and it can be seen that solid waste is a low sulfur fuel. This is clearly shown in Figure 3 where the pounds of sulfur emitted per 10^6 BTUs released is compared to other fuels. This clearly shows that solid waste emits far less SO_2 pollution than either coal or oil.

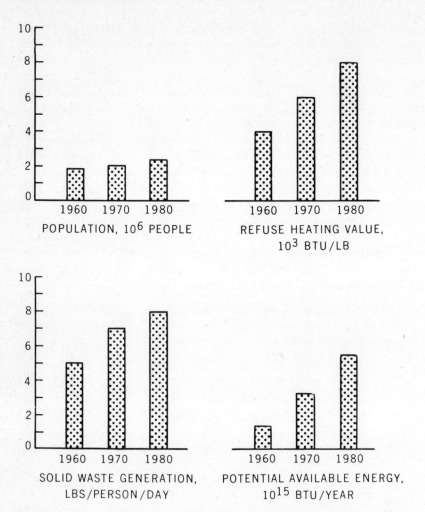

Figure 2 Four factors which make solid waste a potential energy source

A strong case for using municipal solid waste as a fuel can be built when we consider only its chemical composition, low sulfur content, heating value, and logistics of supply without regarding the physical character of the solid waste. Once the heterogeneous nature of solid waste is established the picture changes. We are reminded of solid waste as it is commonly seen, dumped without consideration of recovery in unsightly "sanitary" landfills or burned in the poorly managed, inefficient incinerators in common use in the U.S. today.

In Europe, where fuel is less plentiful than in the U.S., it is common practice to operate incinerators in a clean and efficient manner so as

Table 1 Municipal Solid Waste Components

Component	As Re- ceived-Wet (Wt. %)	Picked- Dried (Wt. %)	BTU/lb. Dry
Paper	48.0	48.7	7,970
Leather; Plastics; Rubber	2.0	2.8	10,000–11,500
Garbage (Food Waste)	16.0	11.1	8,540
Grass; Tree Leaves	9.0	6.9	7,300
Wood	2.0	2.1	8,600
Textiles	1.0	0.7	8,040
Glass, Ceramics, Stones	14.0	16.6	84
Metals	8.0	11.1	742
Total	100.0	100.0	—

Higher Heating Value = 3000–6000 BTU/lb. as received.
Moisture Content = 20–40 Percent.

to convert a significant portion of their solid waste to steam or electricity. It can be done; the technology has been practiced for many years in Europe and could be imported. There are some exceptions, notable is the new Chicago facility where heat recovery is practiced.

The incineration of refuse has failed to meet our needs and the history of incinerators in the U.S. has seen many failures and few successes. Public sentiment and government officials appear to be hardened against the incinerator. The debate raging over the waste disposal

Figure 3 Comparison of solid waste to other premium fuels

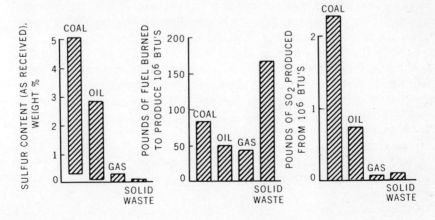

Table 2 Physical and Chemical Characteristics of Municipal Solid Waste

Constituents	Percent by Weight (As Received)
Proximate Analysis	
Moisture	20–40
Volatile Matter	50–65
Fixed Carbon	3–9
Non Combustibles	15–25
Ultimate Analysis	
Moisture	20–40
Carbon	15–30
Oxygen	12–24
Hydrogen	2–5
Nitrogen	0.2–1.0
Sulfur	0.02–0.1
Non Combustibles	15–25
Empirical Formula Solid Waste	$C_{30}H_{48}O_{19}N_{0.5}S_{0.05}$

problem in our nation's capital is an example. In a way, this may prove fortunate. With the background of many unsatisfactory experiences with incinerators, based upon yesterday's technologies and needs, it is easier to provide for future needs with new up-to-date technology.

CONVERSION PROCESSES

Several processes for the utilization of municipal solid waste will be described. These do not represent a comprehensive list but provide typical processes that use the solid waste directly as a fuel or as a new material for the manufacture of liquid or gaseous products.

In order for solid waste to obtain acceptance as a fuel, its physical appearance must be altered. It is not possible to have old pianos, dead cats, used tires, egg cartons, plastic film, and putrescrible organics considered as a solid fuel. It is necessary to convert it to a more desirable form such as a homogeneous solid fuel, homogeneous liquid fuel, or gaseous fuel.

It must be recognized that the necessary condition to render the solid waste useful is size reduction and some blending of feed stocks. Once this has been done a wide range of alternatives are available for utilizing the energy content.

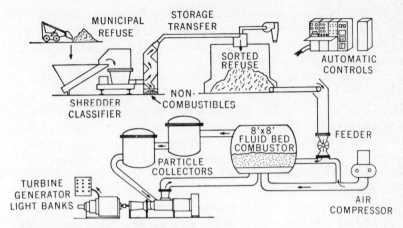

Figure 4 Schematic of the CPU-400 Process

The Union Electric Co. of St. Louis is feeding pulverized refuse directly into a pulverized coal electric generating station. Solid waste will provide about 19% of the energy fired. If this process proves viable, a single power plant of 100 MW would be able to consume about 170 tons of waste per day. The Combustion Power Company of Menlo Park, California, is pioneering an advanced approach to solid waste electrical power generation (CPU-400). A schematic of their system is shown in Figure 4. The solid waste is reduced in size and fed to a pressurized fluidized bed combuster. The hot gases resulting from combustion are cleaned of particulate matter and passed through a gas turbine to produce electricity. A research and development program supported by the Environmental Protection Agency has been underway for four years. Successful development could result in a unit that has the following advantages over conventional incineration for power generation:

1. size of the plant for a given capacity is much smaller
2. moderate size that can be used by small communities
3. large communities will be served by several units located to minimize transportation costs
4. effluent from the plants will meet existing air pollution regulations
5. economic incentives over incinerators

The two processes described above use the solid waste directly to produce electricity. Work on both of these processes is underway. There is concern over the effects of chlorine in the off gases resulting from the combustion of polyvinyl chloride present in the solid waste both as a pollutant and for its corrosive effects on the process itself.

In contrast to those processes that degrade the solid waste to their combustion products, the waste may be processed to yield a gaseous or liquid fuel that retains most of the energy originally available in the solid waste. If it is found necessary to remove contaminants, such as chlorine from the fuel, it should be less costly to develop clean-up processes for homogeneous gases and liquid than directly for solid waste. As is the case with SO_2 removal, it would be expensive to remove contaminants from the products of combustion because of large volumes and low contaminant concentration. The homogeneous gas or liquid produced from the solid waste may be used in a number of processes as a direct substitute for our natural gas and oil.

Two processes will be described for the production of liquid fuel along with a discussion of two processes for gaseous fuel production. The liquid fuel has the advantage of easy transportation and storage, and it need not be used at the same location as where the solid waste is deposited and processed. If the gaseous fuel produced is of pipeline grade it can be easily and inexpensively transported.

The Garrett Research and Development Company of LaVerne, California, the central research organization for the Occidental Petroleum Corporation, has reported a process which converts municipal solid waste into oil, fuel gas, and solid char (Mallan, 1971). The process is essentially one of iron removal, air-classification, shredding, and pyrolysis. The pyrolysis process is essentially one of rapidly heating carbonaceous material in the solid waste in an oxygen-free atmosphere using a proprietary heat-exchange system. The pyrolysis products consist of 40% oil, 35% char, 10% gas, and 15% water. The oil yield is over one barrel of fuel oil with a heating value of 12,000 BTU per lb. per each ton of wet (25% moisture), as received, municipal solid waste processed. The fuel gas fraction can be increased using the same process. Over 80% of the shredded refuse can be converted to a fuel gas with a heating value of 770 BTU/ft^3. A four ton per day pilot plant was scheduled to begin operation in November 1971.

The U.S. Bureau of Mines, Pittsburgh Station, is currently developing a process which can convert municipal solid waste or lignite into a low sulfur fuel oil with a relatively high heating value (*Chemical Engineering*, 1971). The solid waste is allowed to react with carbon monoxide in the presence of sodium carbonate (catalyst) to produce the fuel oil.

The actual process involves feeding the solid waste in a water slurry along with the CO rich gas through a pre-heater and into the stirred reactor at 350°C and 2000 to 4000 psi. The fuel oil yield is 20% to 30% based on the weight of dry solid waste and has a heating value of 15,200 BTU per lb.

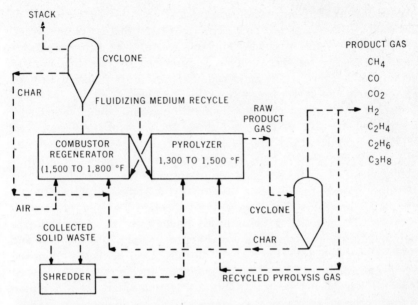

Figure 5 Possible schematic for the West Virginia University Process

Both of these processes convert the solid waste to oil and give yields of about 1.8 barrels per ton of solid waste or about 0.3 pounds of oil per pound of solid waste. The total energy value of the oil is about 50% to 75% of the original solid waste.

There are many processes being developed for coal gasification, and most of these processes could use solid waste as their feed stock. These processes are, however, usually quite large, operate under rather severe conditions of temperature and pressure, and have high capital costs. It is much easier to produce gas from solid waste than from coal using the simple process of pyrolysis. Research work being carried out in our laboratories at West Virginia University has studied the pyrolysis of solid waste in a fluidized bed. The work is sponsored by the Solid Waste Office of Research and Monitoring of the Environmental Protection Agency. The purpose of the investigation is to obtain a high yield of gas from the pyrolysis of municipal waste. It has been demonstrated that fuel gas with a heating value of about 400–500 BTU per cubic foot can be produced that contains over 70% of the original combined energy of the solid waste. Based upon these results a process for conversion of solid waste to fuel gas has been proposed. It is shown in Figure 5. It is similar to the catalytic cracker used in the petroleum industry in the manufacture of high octane fuel. In this case, two fluidized bed reactors are fluidized with an inert solid. The two beds are operated at different temperatures, and there is an exchange of inert solid between the two beds. The right hand bed is the pyrolyzer. Cellulosic wastes are introduced into this reactor where they are pyrolyzed. The energy for the pyrolysis is furnished by the circulation of the inert solid from the left hand bed. This reactor is fluidized with air and combusts the char formed from the pyrolysis. The fuel gas produced from this system contains little nitrogen and could theoretically be upgraded into 1000 BTU gas, a direct replacement for natural gas. For many uses it would not be important if the gas produced had nitrogen present. In this case, a simpler system composed of a single reactor, where both the pyrolysis and combustion take place, is possible.

Another process that is being investigated to convert solid waste to a fuel gas is anaerobic digestion. Anaerobic digestion is a process by which organic material is decomposed biologically in an oxygen-free environment. The produced fuel gas is approximately 70% methane and 30% carbon dioxide. The carbon dioxide may be easily removed leaving essentially pure methane, a premium fuel presently in limited supply. It has been reported (Bohn, 1971) that if all the 1.5 billion tons per year of animal wastes were treated by anaerobic digestion, $10^{13} ft^3$ of methane could be produced. This is nearly half of our current national methane consumption.

So far nothing has been said concerning the economics associated with each of these processes. In order to obtain some feel for the economic potential of energy recovery from the solid waste, let us consider a hypothetical example. Consider a city of 150,000 persons producing 400 tons of waste per day along the east coast called city XYZ. The community has been incinerating their solid waste at a cost of $4 per ton. Recent air pollution regulations have been passed that restrict operation of the local incinerator. This same law also places strict limitations on SO_2 emissions. Figure 6 shows the dollar value of the solid waste as a replacement for the present fossil fuel utilization. A penalty is added to the cost of coal for the removal of SO_2. Without this penalty, the value of solid waste as a replacement fuel would be on the order of $2400 to $3400 per day. The waste would have this as a maximum value. The cost of processing the solid waste must be subtracted. The process development has not progressed far enough to make accurate estimates of the costs, but there are estimates (Mallan, 1971; Rosen et al., 1970), based upon feasibility studies, that show the cost of disposal to range between $2 to $5 per ton. If these are correct, the utilization of solid waste as a fuel has economical potential as well as the potential to conserve our natural resources. In the next section an effort will be made to describe some of the reasons why this energy is not being recovered.

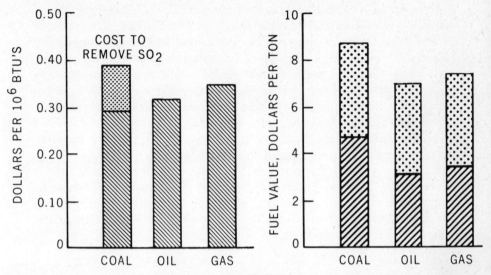

Figure 6 Fuel value of solid waste as a replacement for other premium fuels

BARRIERS TO ENERGY RECOVERY

There are many barriers that stand in the way of widespread recovery of the energy value of wasted solids. Some of them are technical and some economic. Probably the major barrier to effective utilization, however, is the general attitude of industry, elected officials, public works administrators and engineers, and the public.

Although we hear more and more about recycling and recovery from a small segment of our public and see a few civic organizations undertaking recycling programs, there appears to be little real general support. Few of our citizens would consider the slightest inconvenience or cost in order to conserve our resources by the recycling of waste material. This is the age of "do our own thing" with little regard for the consequences of our actions as they effect our future.

One might have hoped to find guidance and leadership in the conversion of the solid waste from among administrators and city engineers given the responsibility for handling this waste. But their general attitude has too often been expressed as:

The collection and disposal of these wastes which, by definition *have no economic value.* . . .

The statement is grammatically correct, but it is misleading and helps to muffle voices calling for recycling. The public is led to believe that the material collected from our homes, the material fed to the incinerator, the material buried in a land fill has no value. During WW II we realized value from this material, and we could today if we would believe in the cause of conservation. We cannot look to the municipalities for leadership in the recovery of valuable resources from the wasted solid. They consider the municipal refuse solely as a problem to be minimized. They do not look at it as a natural resource that is to be utilized for the public good. Where then can we look for answers to conversion of solid waste to useful products?

The manufacturing industry is in the business of taking a material of small value and converting it into a product of greater value. Why can they not convert this waste material into a product of more value? They could if they felt it was sufficiently in their interests. But what is the attitude of industry? It varies, of course, but the fact that it is not heavily involved suggests it has not put forth much effort. One major company that is involved asserts:

Recycling and recovery holds a lot of fascination for politicians and the lay mind. There is a lot of garbage written about garbage.

One reason given by industry for not becoming involved is simply the politics that are involved. Tradition has placed the problem of municipal waste disposal in the hands of local government, and it is considered too hazardous for industry to become entangled in these affairs. Industry certainly has the technical capabiltity to see giant steps made in the recovery of resources from waste but has not taken an active role.

The U.S. Congress passed the Resource Recovery Act of 1970 that charged the then Bureau of Solid Waste Management (presently part of the Environmental Protection Agency) with the responsibility of supporting programs for the recovery of resources and, in particular, energy from our wasted solids. They did not, however, provide the funds necessary to carry out this mission. In spite of the lack of sufficient funds, much of the progress that has been made has been a result of this legislation. Most of the processes described earlier have received government funding in some manner. They lack the necessary funds to adequately support alternative processes, to adequately monitor those they do support, or the technical staff to assess the relative potential of competing conservation processes. The federal government has assumed the leadership role and has achieved some notable successes.

The program of the federal government has provided a valuable boost toward the recovery of energy from solid waste. It has helped to create a change of attitude; it has been able to demonstrate that processes for the recovery of resources are possible. It has seen industry become involved, and more and more people demand that we at least try to recover some of our resources from this material.

If we are to see an effective and efficient system, or systems, for the recovery of energy from our solid waste, industrial participation is absolutely essential. They are the segment of our nation that has the technical capacity to achieve a solution. The municipalities do not have it; the federal government does not have it; the universities do not have it. It must, somehow, be placed on a competitive basis with an economic incentive that will attract industry. It is the competition that will screen the various alternatives and assure that only the more efficient processes are fully developed. However, if active industrial participation is not obtained and the federal government is left to make the decision regarding which process to develop, there is a significant risk that the nation could be shackled with a poor process that could result in setting back several years the chance for the effective recovery of energy from solid waste.

If we are to see efficient utilization of municipal solid waste as an energy resource, there is going to have to be a break in the tradition of the past. No longer is the waste problem going to be left solely to the municipalities. They do not have the technological capabilities or the financial capacity to develop modern processing facilities. No longer can industry shirk its responsibility to apply the technological capabilities to seek solutions that are in the national interest. No longer can the national government fail to provide national policies that will make effective use of our national resources. No longer can the public continue to place their personal convenience and comfort above problems of the environment and conservation of the nation's resources.

There are technical and economic problems that must be resolved. But these appear small in comparison with providing a climate where all of the forces mentioned above can work in harmony toward a solution to energy recovery from wasted materials. This could be a rare opportunity to demonstrate what may be accomplished if we can work together.

CONCLUSIONS

The following conclusions may be drawn from this study:

1. Solid waste utilization would conserve natural resources.
2. The energy available in solid waste is replaceable from the sun's energy.
3. The generation rate of solid energy is increasing, along with our energy needs.
4. Most of the materials discarded by our affluent society are wasted.
5. The wasted solids collected could provide up to 6% of our national energy requirement.
6. The wasted solids are a low sulfur fuel.

7. Not all wasted solids are collected and concentrated in the quantities necessary for energy recovery.

8. Municipal solid wastes are concentrated in areas that desperately need low sulfur fuel.

9. Compared to fossil fuels solid wastes have a potential value of from $6 to $9 per ton.

10. Processes are being developed that should lead to efficient recovery of energy from solid waste.

11. Industrial participation is essential to the development of effective processes.

12. Tradition is probably the largest hurdle to effective recovery of the available energy.

Solid waste cannot furnish a major portion of our energy needs. It represents a small, but not insignificant, portion of our energy requirements. It is, however, a low sulfur fuel, is available in the regions where a fuel crisis is often found, and is a growing and replaceable source of energy. Progress, both technological and political, is being made toward the eventual recovery of this energy resource.

REFERENCES

Bailie, R. C., 1971. Gasification of solid waste materials in fluidized beds. AIChE 69th National Meeting, Cincinnati, Ohio, May.

Bohn, H., 1971. Methane from waste. *Environ. Sci. Technol.* 5, no. 7: 573.

Chemical Engineering, 1971. Carbon monoxide wrings oil from lignite or garbage. vol. 78, no. 24 (Oct.).

Clark, Wilson, 1971. How to harness sunpower and avoid pollution. *Smithsonian* 2, no. 8 (Nov.).

Glaser, Peter E., 1968. Power from the sun: Its future. *Science* no. 162, (Nov.).

Mallan, G. M., 1971. Total recycling process for municipal solid wastes, AIChE 70th National Meeting, Atlantic City, N.J., September.

Mills, G. A. and Harry R. Johnson, 1971. Fuels management in an environage, *Environ. Sci. Technol.* 5, no. 1.

Rosen, B. H., et al., 1970. Economic evaluation of a commercial size refuse pyrolysis plant. Report of the Cities Service Oil Co., Cranbury, N.J., Mar.

DAVID J. ROSE

Controlled Nuclear Fusion: Status and Outlook

The attempt to generate power by controlling nuclear fusion will make an interesting topic for philosophers and historians of science and technology. If such an extravagant statement sounds forced, it is just meant to say at the outset that many factors, not all scientific, and some for the first time, have helped put the state-of-the-art where it is now. We shall try to give some account of these things.

ELEMENTS OF THE PROBLEM

Controlled fusion research has passed through several epochs, the first of which was initiated by four items. First came measurements of reaction energies and rates between hydrogen isotopes and other light elements, which showed that under proper conditions large energy releases would be possible. The well-known laws of single particle physics seemed to show how an assembly of high energy ions and electrons could be confined in magnetic fields long enough to establish the proper conditions. Alert, the radioactive ingredients and by-products of fusion appear to be much less hazardous than those associated with nuclear fission: therefore, fusion reactors would be simpler and safer than fission reactors. Fourth, deuterium is a fusion fuel in plentiful supply—one part in 7000 of ordinary hydrogen; and extraction from ordinary water is not difficult. So matters stood in the early days, say up to 1955. Only the first of these items is necessary to make H-bombs. The combination of all four items captured the imagination of a sizable and very competent fraction of the physics community. The ensuing search for controlled fusion—the ultimate power source— has sometimes taken on a moral character, possibly as a reaction to the darker uses to which nuclear energy had been put. Whatever the reason, the efforts exerted by some might be compared to those of an Everest climber who knew that Prometheus was chained to the top. And a good thing, too, for the 1953 worker didn't see the whole field of plasma physics that lay yet to be discovered between his hopes and their realization. Whether it is a field or a gulf is yet to be discovered, and attempts to cross it during later epochs are briefly accounted below.

The present consensus is that, scientifically speaking, controlled fusion is probably attainable. But if fusion reactors are to be truly practical, there are other requisites: producing large volumes of magnetic field at low cost, minimizing the effects of material damage by high energy neutrons, and so forth. All these are equally essential to success; their natural laws being better understood than those of plasma physics, less room exists either for maneuver or speculation.

These phrases introduce several major topics: how things are now, what is still needed to demonstrate scientific feasibility, what more is needed to make a practical fusion reactor, and how fusion does or does not fit our supposed future requirements.

Several exothermic fusion reactions exist. The reaction of deuterium (D) and tritium (T)

$$D + T \rightarrow {}^4He + n + 17.6 \text{ Mev} \tag{1}$$

is the most attractive, and this discussion is built upon it. The energy is small compared with 200 megaelectron volts per reaction from uranium fission but is more per unit mass. At about 100 kiloelectron volts, the reaction cross section reaches a peak at 5×10^{-28} square meter, which is very large by nuclear standards. Of the 17.6 Mev, 3.5 appears with the ${}^4He$ nucleus, and 14.1 with the neutron.

Many difficulties in the way of developing fusion power can be derived from these simple facts. First, consider the nuclear fuel. Deuterium is almost cost-free, but tritium does not occur in nature and hence must be regenerated with the neutrons from the fusion reaction.

The worst problem is presented by the nature of the reaction itself, because the particles must have (about) 10 kev energy or more so that the D and T nuclei can overcome their mutual electrostatic repulsion and fuse. Unfortunately, the cross section for scattering via this repulsion considerably exceeds the fusion cross section at such energies; hence the particles scatter each other several times before reacting. Thus it follows that the fuel will be a randomized collection of ions whose average energy must exceed 10 kev. In conventional terms, this is a gas at a temperature exceeding 10^8 degrees Kelvin. In fact, it will be a fully ionized plasma of D+ and T+ ions containing an equal total density of electrons to make the medium macroscopically neutral.

As has been implied, the principal difficulty comes in confining this plasma. A D-T nuclear explosive device stays together long enough—less than 10^{-7} seconds—by inertia alone for the components to react. In the process, the ^{4}He nuclei (and to some extent the neutrons) slow down in the unreacted material and heat it to an "ignition" temperature; transient pressure is millions of atmospheres. For a slower, controlled reaction, the pressure must be something that real structures can withstand; systems that we visualize will have dimensons of the order of 1 to 10 meters, and therefore pressures exceeding (say) a few hundred atmospheres are hardly believable. This restriction, plus specification of temperature already made, determines the density of the ions. Depending on the arrangement, desired D + T ion density turns out to be 10^{20} to 10^{22} m^{-3}, some 7 to 9 orders of magnitude below solid densities, and 4 to 6 orders of magnitude below that in the air around us. Required confinement time for a useful fraction of the nuclear fuel to react is 0.01 to 1 second. The most important parameter is the product of the density by the time, which should be 10^{20} sec m^{-3} or more—the so-called Lawson criterion. Total reacting nuclear mass at any one time would be only about one gram, even in a system that operates continuously at several thousand megawatts. All this is remote from any explosive regime.

PRESENT SCIENTIFIC PROGRAM

We will not review in depth the voluminous plasma physics underlying the schemes by which the plasma is hoped to be confined; but some acquaintance is necessary for what follows. The main schemes being developed so far involve use of large volumes of high magnetic fields. Plasma ions and electrons are hindered by magnetic forces from moving across the direction of magnetic fields, but can spiral along the field lines, as in Figure 1. Thus (naively), confinement in the two directions perpendicular to the field direction is achieved, and one might have to worry only about confinement along the field direction.

From these simple thoughts arose in the first epoch two largely separate categories of device (Bishop, 1958; Alexander, 1970; Fowler and Post, 1966). In Figure 2, field lines are curved to form a closed toroidal system; there is no escape except across field lines, and devices of this generic type are called closed systems. In the other generic type of Figure 3, ions (and electrons) are reflected by increasing magnetic fields at each end. Here, an additional mechanism is required: each ion moving along a magnetic field line has fixed total kinetic energy U—at least until it interacts with the other ions and electrons in the system, or undergoes fusion. The total energy U can be thought of as being

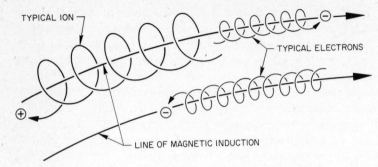

Figure 1   Orbits of ions and electrons in a magnetic field

composed of two parts, and energy $U\perp$ of gyrating motion perpendicular to the field line, and a part $U_{\parallel}$ of motion along the field line. That is

$$U = U\perp + U_{\parallel} \tag{2}$$

Now it can be shown (Rose and Clark, 1965) that the magnitude of the perpendicular component $U\perp$ is proportional to the magnitude B of the magnetic field; that is

$$U\perp = \mu B \tag{3}$$

where μ is a constant (called the magnetic moment) for each particle, depending on details of its orbit. From this we find

$$U_{\parallel} = U - \mu B \tag{4}$$

Figure 2   Toroidal magnetic field B_ϕ made by poloidal electric currents I_θ

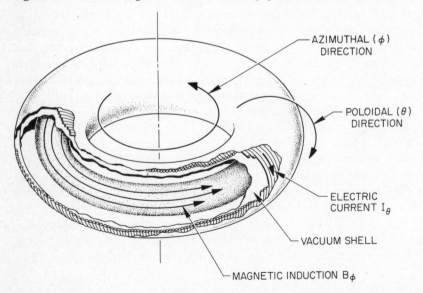

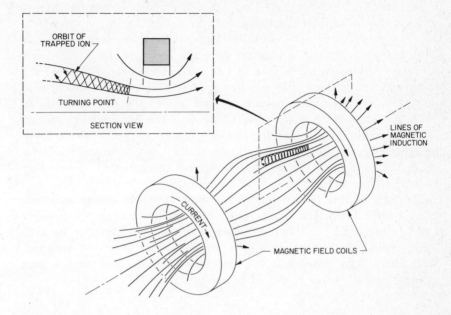

Figure 3 Magnetic mirror particle (and plasma) confinement configuration

The consequence of Eq. 4 is straightforward—if the field B becomes high enough in the ends of the device shown in Figure 3, then μB rises to equal U itself, and no energy $U\perp$ is left for parallel motion. The particle must be "reflected" from these high field regions, hence contained in the center part. The device is appropriately called a magnetic mirror. (This scheme is what confines high energy ions and electrons in the Earth's van Allen radiation belts. Particles are reflected above the ionosphere near the north and south magnetic poles, and follow the lines of weak field far from the earth in between.)

A difficulty of these "open-ended" systems of Figure 3 is just that— open ends. An ion or electron whose orbits happen to lie almost along the field direction in the middle of the device has a low value of the magnetic moment. Then the maximum field B at the mirrors is insufficient to reflect the particle, and it escapes out one end. Coulomb interactions continually scatter particles in to such directions; hence magnetic mirrors are inherently leaky, even if no worse calamities befall.

In each case, the confining field might typically have a maximum strength of 8 to 10 tesla (1 tesla = 1 weber/m², or 10,000 gauss. Earth's field at the surface is about 0.5 gauss.) and an equivalent magnetic pressure $B^2/2\mu_o$ in meter-kilogram-second units) of 300 atmospheres.

The difficulty with all these truly ethereal schemes is that the plasma turns out to be unstably confined, because a number of electric effects which are negligible for a few isolated particles but important if a large assembly (that is, a plasma) were not included. Thus ended the first epoch of fusion research, a sort of age of innocence. For either the closed or open systems of Figure 2 or 3, some field lines necessarily bow outward away from the plasma; at such places the plasma tends to develop uncontrolled aneurisms. Modifying the basic configurations (and increasing its cost and complexity substantially) will reduce these unstable growths, but it seems certain that a weak turbulence will remain. As a result, plasma could diffuse toward the surrounding vacuum walls and out the ends at a high rate.

The idea of diffusion is useful for illustrating the situation in the present second epoch of fusion research. If the plasma internal motions can be described by a diffusion theory (there is some doubt about this, which we ignore here), then a diffusion coefficient D can be assigned. The theory then states that the confinement time τ_c in (say) a long cylinder of wall radius r_w should be about

$$\tau_c = r_w^2/6D \tag{5}$$

For long τ_c, we desire small diffusion, but even more importantly large systems. Present custom (Taylor, 1962) has it that the diffusion coefficient is likely to be some small fraction of the Bohm value D_B for a fully turbulent plasma, where

$$D_B = \frac{kT_e}{e} \cdot \frac{1}{16B} \tag{6}$$

Here, (kT_e/e) is the electron temperature measured in electron volts. Then according to this rubric, we have

$$D = D_B/A \tag{7}$$

where the dimensionless factor A represents confinement quality, measured in "Bohm times." If $A = 1$, the plasma would be lost by diffusion with a co-efficient equal to D_B. For adequate fusion system confinement, it turns out that we must have $A \gtrsim 100$ at least, the precise number depending upon the arrangement (Rose, 1969).

It is both encouraging and salutory to see where present experimental devices are in relation to these goals. There are many such, but in this summary one example must suffice. The Tokamak, one of the most promising devices today (Artsimovitch, Bobrovsky, et al., 1969), is an easy extension of Figure 2, developed first at the Kurchatov Institute in Moscow, now also appearing in various guises at several plasma laboratories in the United States. Figure 4 shows the arrange-

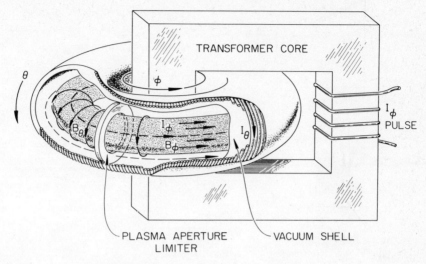

Figure 4   The Tokamak plasma confinement scheme

ment: the strong azimuthal field B_ϕ remains as before; but now the toroidal plasma is itself also the secondary loop of a transformer, which accomplishes two additional purposes. First, a strong current pulse on the primary winding ionizes the gas and generates a secondary plasma current I_ϕ; that current heats the plasma by inducing weak dissipative turbulence—hopefully just enough to heat it but not lose it (Fig. 4). Second, the current I_ϕ produces a new poloidal magnetic field B_θ as shown; the two fields combined, reminiscent of the crossed plies of a tire tread, make up the confining structure. Analysis shows that the plasma should be stable against ordinary hydromagnetic instabilities in the magnetic well so formed. The remaining higher order modes might be too weak to cause excessive diffusion. One penalty for these improvements is abandonment of true steady-state operation, for the device must now be run in long pulses—vide the transformer.

At this time, hopes that a Tokamak device will establish the scientific feasibility of fusion reactors are high. The largest device operating ("T-3" at Kurchatov) has a major diameter of 2.0 m, the minor plasma diameter is about 0.3 m, the maximum field B_ϕ is 3.5 tesla, and the current I_ϕ is 10^5 amperes. For these efforts, the results (Artsimovitch, Anashin, et al., 1969) are: plasma density is $3 \times 10^{19}/m^3$, confinement time τ_c is 0.03 seconds, the electron temperature is > 1 kev, and the ion temperature is 0.5 kev. Each of these numbers (which has been measured both by the USSR and a visiting team from the United Kingdom) is about a factor of 10 too low, but very good by recent standards;

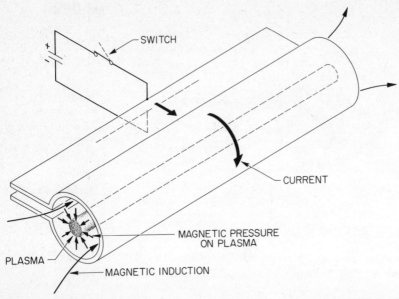

Figure 5   Pulsed plasma heating and confinement scheme (so-called θ-pinch)

and there is more to the story. From Eqs. 5 to 7, we calculate $A \approx 80$; that is, the confinement time of 0.03 second is some 80 times as long as turbulent Bohm diffusion would predict. This bespeaks a fairly quiescent plasma, almost good enough (in these peculiar terms) for a fusion reactor. A respectably optimistic expert could argue that only the small size and relatively low magnetic field prevent the plasma from lasting an adequate number of seconds. Exploring whether larger or higher field devices give a closer approach to fusion reactor parameters is now an exciting activity; the next generation of experiments should tell much.

Analogous descriptions might be made about some magnetic mirror experiments (the so-called 2X experiment at the Lawrence Radiation Laboratory, Livermore, California, for instance (Fowler, 1969; Coensyen et al., 1969), or fast shock-heated plasmas—Scylla at Los Alamos, for example (Little et al., 1969; Beach et al., 1969). This last device is shown very schematically in Figure 5. The capacitor discharge through the single-turn coil generates a rapid-rising strong magnetic field ($< 10^{-6}$ second, 15 tesla). The field acts as a radical piston, compressing an initially cool plasma into a hot, dense one. In each of these various schemes, the combinations of density, temperature, and confinement time differ. For the Scylla experiment, we find densities up to 5×10^{22} m^{-3}, and temperature ≈ 5 kev, which are nearly satisfactory for fusion; but $\tau_c \approx 10^{-5}$ second is very short: plasma squirts out the open ends of the device. A longer one (Scyllac, 10 m) is being built to reduce these end effects.

GENERAL TECHNOLOGICAL FEASIBILITY

Divinations from plasma physics may permit or deny the possibility of useful power from controlled fusion, but they cannot guarantee it. Some applied problems that are substantially independent of the particular geometric model are:

1) Plasma conditions in imagined practical devices, such as ion and electron temperatures, the fraction of fuel burned up per pass through the reactor, and radiation from the plasma surface. This might be called plasma engineering.
2) Regenerating tritium (for a D-T reactor) in a surrounding moderator-blanket by means of the 14.1-Mev neutrons.
3) Heat deposition, temperature of the moderator and vacuum wall, and heat removal.
4) Providing large quantities of high magnetic field and structure to withstand high stress.
5) Radiation damage by the 14.1-Mev neutrons, the consequences of which may be frequent and expensive replacement of much of the structure.
6) Size and cost, which are implicit in many of the above. Other problems are model-dependent; some device concepts seem to require additional developments. The list is long.

Most of the engineering-type problems that are model-independent can be described with the aid of Figure 6, which shows a stylized fusion reactor as a series of cylinders. The main confining magnetic field is into (or out of) the paper; whether the cylinder is the center section of a stabilized mirror or is wrapped into a torus need not concern us here. The fusion plasma occupies the evacuated center, is surrounded by a neutron-moderating blanket, and, at large radius, by a set of magnetic field coils. Here now are summary remarks on the problems listed above, generally slanted to a steady-state (or quasi-steady-state) device (Rose, 1969; Hall and Maple, 1970).

1) The Plasma. How is the plasma heated? What are the equilibrium temperatures and other parameters? The confinement being imperfect, we imagine plasma fuel continually being lost from the ends or sides into some suitable pump, hence also being replaced by some injection process into the center. Thus, the plasma continues in existence, but each ion or electron remains confined only for the period τ_c discussed before. Helium nuclei born in fusion reactions are also trapped for about

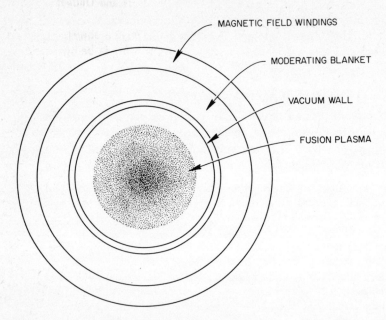

Figure 6 Schematic controlled fusion reactor

τ_c, and deliver much (possibly all) of their 3.5-Mev energy to the plasma. Thus, the plasma is at least partly heated by its own reaction. For some fixed τ_c, then, a certain through-put of plasma is needed to keep up its density; consequently, a certain calculable fraction f_b of the fuel will be burned per pass through the device; and the helium from the reaction heats electrons and ions (unequally) to temperatures T_e and T_i, respectively. As τ_c is raised, then f_b, T_e, and T_i also go up; the fuel is confined better and is not diluted by so much unreacted through-put. Fractional burnup f_b is a more useful display criterion than is τ_c. Difficulties of replenishing the fusion plasma seem to limit us to $f_b \gtrsim 0.02$; $f_b > 0.1$ would cause too high plasma temperatures and also demand unimaginably good confinement.

With some rather restrictive assumptions, these things can be calculated. Figure 7 shows the expected rise of electron and ion temperatures with increasing fractional burnup, for typical conditions expected in a fusion reactor. At high f_b, electron temperature falls below that of the ions. The reasons for this are that energetic electrons radiate energy, and that the ^{4}He nuclei tend to heat the ions preferentially, if the electron temperature exceeds about 33 kev.

Are these temperatures (once established by some startup scheme) high enough, or must more energy be added? This question lies at the

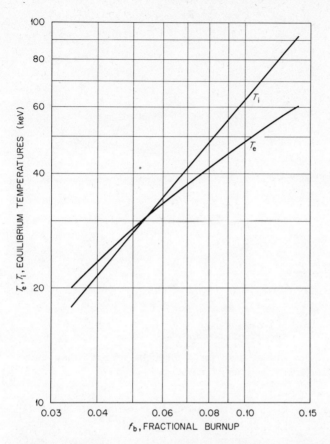

Figure 7 Electron and ion average energies expressed as temperatures T_e and T_i, respectively, in projected fusion reactors, as a function of the fraction f_b of injected nuclear fuel that is consumed

heart of determining energy balance in a fusion reactor. At a given plasma pressure, the highest fusion reaction rate per unit volume occurs at temperatures of 15 to 20 kev. Then Figure 7 appears to show ample heating if only $f_b \gtrsim 0.03$. For toroidal systems, this may be satisfactory, but an additional problem appears for open-ended systems (mirrors): the ions scatter out of the ends intolerably rapidly unless the ion temperature is very high, perhaps 100 kev or more.

For mirrors, heating the ions (probably by injecting them into the plasma at high energy) appears to be a necessary but expensive step. The expense arises both in additional equipment and in energy. Most of the energy from a D-T fusion reactor will appear as heat, which can be converted to electricity with (at most) about 50% efficiency.

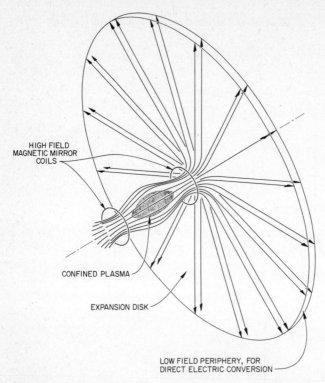

HIGH FIELD
MAGNETIC MIRROR
COILS

CONFINED PLASMA

EXPANSION DISK

LOW FIELD PERIPHERY, FOR
DIRECT ELECTRIC CONVERSION

Figure 8 Magnetic field configuration for a magnetic mirror fusion device with direct electrostatic energy recovery (after Post et al., 1970)

Then using large amounts of electric power to inject ions could make the system unfeasible.

These objections are serious enough so that a very different energy cycle is being investigated for mirrors (Post et al., 1970). The field lines of such a device are shown ethereally in Figure 8. Plasma escaping through the mirror (only one end is shown) is expanded radially to the periphery of a large disk, where the density is so low that electrostatic direct energy conversion can recover the plasma energy with high efficiency. This energy is used (also with high efficiency) to reinject ions. The scheme will not work well with a D-T fusion cycle, but a D-^{3}He cycle which produces charged particle reaction products almost entirely might be better. Such a cycle requires ion energies of several hundred kilovolts, a factor of 10 higher than for a D-T cycle. If the idea works, it would indeed make a virtue of a necessity; but the additional difficulties seem immense, and the outcome is problematical. Nevertheless, it may represent an important hope for the entire class of open-ended fusion machines.

A major difficulty with all these calculations is that they are still nebulous. The hidden assumptions may be unrealistic in serious ways.

For example, how are the energy exchange rates inside the plasma affected by the presence of weak turbulence? No one knows. Will the curves of Figure 7 be affected by inclusion of space charge effects? A subfield of fusion plasma engineering, for lack of a better phrase, needs developing before a fusion reactor can be sensibly designed.

2. *Tritium Regeneration.*   For a D-T reactor, tritium must be regenerated; the two lithium reactions

$$^7\text{Li} + \text{fast neutron} \rightarrow \text{T} + {}^4\text{He} + \text{slow neutron} - 2.5 \text{ Mev} \qquad (8)$$
$$^6\text{Li} + \text{slow neutron} \rightarrow \text{T} + {}^4\text{He} + 4.8 \text{ Mev} \qquad (9)$$

are essential and seem adequate.

The general idea in Figure 6 is, then, to make the vacuum wall and blanket supporting structure of thin section refractory metal. Within it, there would be liquid lithium or a lithium salt coolant, plus an artfully disposed neutron moderator (probably partly of graphite). Leading choice for metals is niobium in that it can be formed and welded, retains its strength at 1000°C, and is transparent to tritium. This transparency helps in two ways: tritium generated in the lithium-bearing coolant is not trapped in the metal; and tritium can be recovered by diffusion through thin section walls into evacuated recovery regions. Some additional neutrons also come from the niobium via (n, 2n) reactions, but in this particular respect molybdenum would be a better material.

Liquid lithium cooling has the advantages of high heat transfer, few or no unfavorable competing neutron reactions; main disadvantage is its high electric conductivity, which makes it hard to pump through high magnetic fields—just how hard is not well enough known. In regions near the vacuum wall where the high lithium flow rate might cause excessive pumping loss, a nonconducting molten salt can be used. The likeliest candidate is Li_2BeF_4; the main penalty for its use is the presence of fluorine, which slows down energetic neutrons unprofitably, hence inhibiting the beneficial ^{7}Li reaction of Eq. 8. That is, using Li_2BeF_4 makes it harder to regenerate enough tritium.

However, with either of these schemes or a combination of them, tritium regeneration seems adequately assured. Calculations with semi-realistic combinations of vacuum wall and blanket show that something between 1.1 and 1.5 tritons can be regenerated per neutron incident on the vacuum wall (Steiner, 1968; Blow et al., in Hall and Maple, 1970). Because one triton is used up per neutron generated, we have in fact a tritium breeder reactor, using the raw materials deuterium and lithium. This view of fusion as compared to nuclear fission breeder reactors has not been much emphasized in the past.

In addition to this favorable breeding ratio, present estimates put the tritium inventory in a fusion reactor at only a few weeks' supply —maybe less (Fraas, 1968). Thus, the tritium fuel doubling time in a fusion reactor might be much less than one year. Doubling time is an important measure of how quickly new reactors could be built (that is, fueled) either to match expanding power demands or to take over from a prior power-generating scheme. This short doubling time for fusion is in marked and favorable contrast to the situation with fission breeder reactors, where the doubling time tends to be uncomfortably long (nearly 20 years in some designs). Here is one of the predicted large advantages for fusion.

Approximate size of the fusion reactor we have in mind comes directly from these considerations. Fairly simple nuclear calculations establish that the blanket plus a radiation shield (not shown) to protect the outer windings must be 1.2 to 2.0 m thick. This substantial thickness implies not only substantial blanket cost, but also very high magnetic field cost, to energize such a large volume. The only way to make the system pay is to have it generate a great deal of power; but nearly all this power must pass from the plasma into or through the vacuum wall. Engineering limits of power density and heat transfer then dictate large plasma and vacuum wall radii as well—between approximately 1 and 4 m. Then overall size will be large, and total power will be high—almost certainly more than 1000 megawatts (electric) and perhaps 5000 megawatts.

3) Heat Deposition and the Vacuum Wall.   Energy is deposited in the vacuum wall facing the plasma, mainly from three sources: (a) some of the fusion neutrons suffer inelastic collisions as they pass through; (b) gamma rays from deeper inside the blanket shine onto the back side; (c) all electromagnetic radiation from the plasma is absorbed there. The plasma itself makes no additional load, being imagined to be pumped out elsewhere. The three sources may constitute 10% to 20% of the total reactor power. This is a modest fraction; but the vacuum wall region is thin, and heat deposition (and removal) per unit volume determines the power capability of the whole system. Here is a disadvantage of fusion systems compared to fission reactors; in the latter the energy is more nearly produced throughout the reactor volume and all must not pass through one critical section.

From these considerations, there is a total power assignment in the reactor of not more than 15 Mw per square meter of vacuum wall— say 10 Mw/m^2 being 14-Mev fusion neutrons passing through, and the rest consisting of plasma radiation and neutron captures in ^6Li. Some (Werner, in Hall and Maple, 1970), imagine substantially higher energy fluxes to be possible, with the use of heatpipe walls—about 30 to 40 Mw/m^2; but the design poses many problems. Even at 15 Mw/m^2, total reactor power is very high, as stated before. If the vacuum wall radius is only 2 m, the system of Figure 6 produces 140 Mw of heat per lineal meter (into the paper) of cylinder. If it is wrapped into a torus, the major diameter can hardly be less than 20 m. Total power of such a device would be 12,000 Mw thermal, or 5000 to 6000 electric, several times that of the largest plant now existing.

One possible way (Fraas and Pease, pers. comm.) out of this and some other difficulties is to run the reactor at substantially lower thermal stress—at $\sim$ 2 Mw/m^2. Total power is conveniently less; and because the plasma density is reduced, so is the magnetic field and the cost of it. Neutron damage is also ameliorated. Whether this option increases the cost per unit of power excessively has not yet been estimated.

The vacuum wall must support approximately a pressure of 1 atm, which is no small task for a thin-section material in such large sizes. However, preliminary designs indicate that a structure built up in depth of thin sheets (the same principle as in corrugated cardboard boxes) will have the necessary strength, and contain proper passages for coolant flow (Fraas, in Hall and Maple, 1970).

4) *Magnetic Field Windings.*   Generating even 15 tesla (150,000 gauss) continuously is not the problem; superconducting coils do so routinely at low cost, a dramatic improvement from state-of-the-art ten years ago. The problem is size: a simple solenoid generating 15 tesla has a magnetic bursting force of 900 atm on its windings. In comparison, contemporary fission reactor pressure vessels are smaller than we imagine here, and are limited to some 40 atm operating pressure. To make matters worse, the magnetic field is not a simple solenoidal one, and stresses arise that cannot be held in simple hoop tension. To be sure, no nuclear excursion impends if the coils fail structurally, but failure would still be an economic calamity. Perhaps also 15 tesla is not required, but no assurance now exists.

Almost all conceptions involve superconducting coils at 4°K, or at least cryogenically cooled at 10° to 20°K. This is the reason for placing them outside the blanket, outside a radiation shield, otherwise the refrigeration problem would be intolerable. To make a reinforcing structure for operation at such a temperature, with size and stress loads described, is a task yet to be fully contemplated. Titanium is very strong at such low temperatures; but it is also very brittle—as are most other materials under those conditions.

5) *Neutron Damage.*   This is a very serious problem, for either a fission or fusion reactor. In one way, fusion appears at a substantial

disadvantage, as follows. One fission reaction produces 200 Mev and about 2.5 neutrons, each with no more than about 2 Mev. One fussion reaction produces 17.6 Mev, of which 14.1 Mev appears in one high energy neutron. Thus, the "energetic neutrons/watt" is an order of magnitude higher in fusion than in fission, and the structural damage caused by these neutrons is correspondingly high. For the high power levels discussed in the preceding examples, every metal atom in the vacuum wall would be displaced almost once per day (Steiner, in Hall and Maple, 1970). Many of these displacements anneal out of the high operating temperature; but, even with the delicate choice of materials, design, and temperature, long-term integrity of the vacuum wall against neutron damage will be a major problem facing fusion power development.

In another way of looking at the problem, fusion has an advantage. The damaging neutron flux in this high power fusion reactor is predicted to be about $10^{15}/cm^2$-sec; but in reference designs for liquid metal fast breeder fission reactors, it will be an order of magnitude higher. We see here a principle of conservation of wretchedness—the fast breeder fuel elements and perhaps the components will require frequent replacement, at substantial expense.

For fusion, this problem translates into the problem of either protecting the vacuum wall (via low power?) or replacing it. The cost of either of these options may be high; unanswered questions are whether the vacuum wall can be replaced at a cost small compared with the total reactor cost and how often replacement will be required.

Compounding the problem are the facts that probable fusion reactor conditions and materials are not in the fission breeder range of interest. Moreover, no source of 14-Mev neutrons (to test possible arrangements) now existing is intense enough—by a factor of ≈ 1000.

Within the framework of fusion systems envisaged here, this damage problem cannot be circumvented, cannot be well predicted on the basis of present knowledge, and affects the feasibility of every fusion reactor scheme.

6) *Size and Cost.*   Size is large for lowest power cost, as shown earlier. However, over many decades unit size has increased by a factor of two to three each ten years. Thus, 10,000 Mw thermal is liable to be quite acceptable before 2000, when fusion might, with good fortune, come into its own.

Cost per thermal kilowatt of capacity makes a reasonable basis for comparison with other generating systems. Components stylized in Figure 6 are equivalent to the core of a nuclear fission reactor, without some of the nuclear ancillaries (and without any of the turbines and generators of a power station). No definite cost can yet be given for what is shown there; too much is still uncertain. However, outside estimates have been made that the cost might run somewhere between 6 and 20 1970 dollars per thermal kilowatt (Rose, 1969). If neutron damage does not require too frequent replacement of the structure, the whole cost range is interesting, and the lower limit is uncontestably attractive.

Such costs warrant continuing development, but they are very perishable commodities, depending on the imperfect and changing state-of-the-art. Designs, costs, trends, and comparisons must be continually reassessed.

MODEL-DEPENDENT PROBLEMS

What of the host of model-dependent problems, more specific than those hitherto listed? I mention just three, to show their kind and importance.

1) Fuel Injection into Closed Toroidal Systems. Plasma is lost by diffusing toward the vacuum wall and then being absorbed (no mean task, and not well understood) at specific peripheral regions. Implicit in this statement is that something replenishes the plasma at or near the middle (if the device runs on anything like a steady-state basis). Ionized particles will not move across the confining field, so neutral ones must be somehow injected. The trouble now is that the energy flux (of hot electrons) in the plasma is about 10^{14} watt/m^2, some 10^3 times that of the strongest electron beam today. Lifetime of a neutral atom or a small cluster of atoms against being ionized in this hostile environment is about 10^{-7} second; upper limit on injected atom velocity is about 10^6 m/sec; otherwise the plasma energy balance is upset. Then the atom penetrates perhaps 0.1 m, a negligible fraction of the way in.

An alternative scheme is to inject pellets so large that they shield themselves by ablation on the way in (as a reentry vehicle into the atmosphere from a space flight). Calculation of what happens here—for example, whether the pellet must be so large that it chokes the fusion reaction—is much more difficult than calculating the fate of atmospheric reentry bodies, and not much has been done (Rose, 1968).

2) Direct Energy Conversion for Open Systems. The necessity for high energy injection and recovery directly as electricity was mentioned in the discussion related to Figure 8. What cannot be illustrated well is that the diameter of the dislike expansion region may be 100 times the diameter of the mirror confinement region. Can such a structure (albeit with low magnetic field) be built cheaply enough? Can plasma stability and individual particle orbits be controlled well enough throughout this immense region? No one knows.

3) Fast-Pulsed Systems. The scheme of Figure 5 has advantages of automatic plasma heating, apparently good stability against radial excursions, and some others. But several perplexing complications are as follows. (a) The system requires a substantial amount of stored energy to be delivered in about 10^{-6} second to the coil. At present this is done by capacitors, perhaps at a cost of \$100,000 per megajoule. Some cost reduction is clearly possible, but much is necessary. (b) The fast pulse requires that the magnet coil be next to the plasma in that it forms the vacuum wall. Then the coil must have high strength at high temperature. Electric losses in this coil reduce power output from the system. The coil also slows down and absorbs neutrons, and this process decreases the tritium yield (Bell et al., in Hall and Maple, 1970; AEC, 1969). (c) Pulsed operation at (approximately) 900 atm pressure on a microsecond basis exacerbates problems of mechanical stress failure; yet more reinforcing structure imperils the tritium breeding even more.

FEARLESS FORECAST

To assess the relative merits of many approaches to controlled fusion is a difficult task, and disputatious. But some sort of perspective must be developed from time to time. What follows is partly opinion, partly fact; it is no one's policy but my own.

Figure 9 helps to focus and confine the discussion. In the middle is a level of achievement called Scientific Feasibility; a density-time product of 10^{20} sec/m^3 or more, and true thermonuclear temperature— say 15 kev or more, depending on the system envisaged. Whether the device looks like any eventual fusion reactor is immaterial in this context. This level of accomplishment would be crudely the analog of building the Stagg Field fission reactor in 1942; the physics is permissive, but engineering and economics are yet to come. Figure 9 has no absolute scale, but shows where each present scheme is presently situated—all are now below the feasibility waterline. Closest is the Tokamak, but the figure shows two gaps yet to be crossed. These gaps are that it is not yet known whether scaling to larger size really will work (as described earlier) or whether the ions can actually be heated enough in the device, via weak turbulence or some other means. To put some calibrating point on all this, I will bet a modest amount of even money on success of the Tokamak in the next few years.

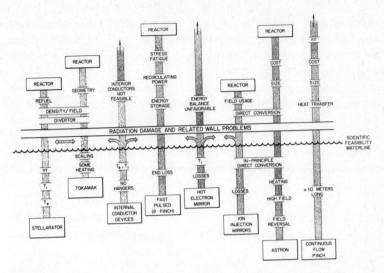

Figure 9 Various paths to successful controlled fusion, with difficulties

The stellarator is a related steady-state device, where the toroidal configuration is stabilized not by induced plasma currents (as the Tokamak), but by added helical windings on the periphery of the torus. The big advantage is steady-state operation. The main disadvantage is that a field configuration made this way seems to give poorer confinement. Thus the density-time product ($n\tau$ in the figure) needs more substantial improvement, and in addition both the ion temperature (T_i) and the electron temperature (T_e) will be harder to raise (Rothman et al., 1969). The stellarator lies significantly below the Tokamak at present.

Some toroidal confinement schemes require solid conductors totally surrounded by plasma. The so-called multipoles at the General Atomic Corporation and at the University of Wisconsin, and the spherator at the Princeton Plasma Physics Laboratory are examples (Yoshikawa et al., 1969). These internal conductors can be (and are) made superconducting, so true levitation without supports or hangers is possible and has in fact been achieved. On the other hand, no large levitated experiment has yet been performed at high enough field. Thus in the third column of Figure 9 we see the need to operate without hangers, and to raise both T_e and T_i by some plasma heating schemes yet to be fully developed.

Next in the figure comes the fast-pulsed devices, as shown in Figure 5. Whether the side losses are now small and whether just reducing end losses will give satisfactory confinement are still questions, but the device should be given the benefit of the doubt. One estimate is that the device needs to be 2 km long if linear and the ends are not stopped up (how?); also if wrapped into a torus, new and unresolved questions of plasma stability enter.

All open-ended mirrors suffer from high loss from the ends, and schemes to reduce these losses (by applying high frequency power at the mirrors, for example) seem not to be very effective. (For an optimistic view, see Ware and Faulkner, 1969.) Heating both ions and electrons adequately is an additional problem. The "hot electron mirror" scheme uses large amounts of microwave power to produce an exceedingly dense hot electron plasma, with apparently fair confinement, at least (Dandl et al., 1969). Ions might be heated (T_i in Fig. 9) by injecting high energy neutral atoms into this "seed plasma." The chances of this scheme making a scientifically feasible fusion plasma are at least fair.

Ion injection mirrors, when the plasma is not substantially aided by hot electrons, face more difficulty. The losses are high; and as discussed above, it seems that the high losses will require as part of the "in-principle" solution the development of "in-principle" direct energy conversion (see again Fig. 8 and the accompanying discussion).

The Astron at Lawrence Radiation Laboratory is interesting, but hard to describe (Fig. 10). It starts out generically as a mirror (Fig. 3); but instead of confining a plasma directly there, the aim is to confine a ring of relativistic energy electrons (relativistic protons in a full-scale reactor). This is called an E layer; if dense enough, its diamagnetism actually reverses the magnetic field and sets up a new configuration of closed magnetic field lines; a torus inside the mirror. This configuration holds the fusion plasma. So far, a modest diamagnetic reduction (and no reversal) of a low field experiment has been achieved (Beal et al., 1969). True field reversal in a larger, high field device will be needed to set up the desired magnetic configuration. Beyond that, how the plasma is to be heated is a problem; and high end-losses may also require direct energy conversion.

The continuous-flow pinch is favored in some quarters, particularly in the USSR. The idea stems from the discovery that plasma can be focused into a small, very high density ($10^{25}/m^3$?), high temperature (several kilo-electron volts) plasma thread a few millimeters long, at the end of a coaxial plasma gun. This is the so-called plasma focus, which is a copious source of fusion neutrons during the time scale of its pulsed operation, about 10^{-6} second. (For an example of the state-of-the-art, see Mather et al., 1969.) Can this very dynamic object be formed and preserved on some more steady-state basis, and spun out from the end of the plasma gun, as a thread from a spinnerette?

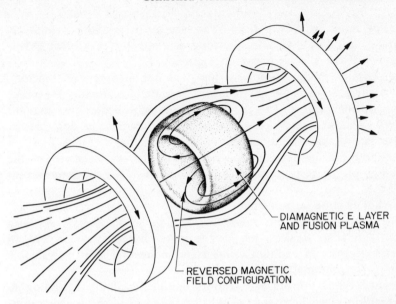

DIAMAGNETIC E LAYER
AND FUSION PLASMA

REVERSED MAGNETIC
FIELD CONFIGURATION

Figure 10 The Astron configuration for obtaining controlled fusion, which aims to generate a strongly diamagnetic region inside a conventional magnetic mirror

No one knows what all the problems are, so we arbitrarily define scientific feasibility as the production of a 10-m thread.

These activities below the waterline of Figure 9 have taken nearly all of the more than $1 billion spent around the world on fusion up to now. But how do things look for making a reactor? Above the line appear many of the problems discussed earlier. Damage to the structure by high energy neutrons may render the whole idea uneconomic, as discussed before. But besides this, the various schemes have different relative merit above and below the waterline.

Tokamaks no longer look quite so attractive. Special plasma pumps called divertors have been developed for stellarators, seen necessary for Tokamaks also (where access is more difficult), but must be vastly increased in effectiveness. Plasma stability considerations may demand that the plasma density be uncomfortably low, or the field uncomfortably high—15 to 20 tesla, or more? (Golovin, in Hall and Maple, 1970; Gibson, in Hall and Maple, 1970.) Also, the geometry, inherently pulsed nature, and necessarily large size of the thing are hard to work with.

Some of these problems appear with the stellarator too, but with reduced intensity. Steady-state operation is easier; the additional refueling problem may be no more than moderately serious. Thus, the stellarator tends to look better, *if* we are given scientific feasibility. Stellarator and Tokamak scientific programs support each other extensively, hence the joining arrow on Figure 9.

The internal conductor devices just will not make fusion reactors, because there is no way of cooling a levitated conductor, especially inside a fusion plasma. This is well understood; no one ever thought otherwise; these experiments are designed specifically for plasma physics and to shed scientific light on other schemes.

The theta pinches have very severe problems, as discussed in the last words of the section on fast-pulsed systems. The outcome is pessimistic, as Figure 9 shows.

Pure hot electron mirrors appear unfeasible for fusion from an energy-balance point of view, but again that is a personal opinion. As with internal conductor devices, the idea is to reach the waterline, not an economic reactor. In addition, *some* electron heating may be valuable for more conventional mirrors.

If conventional mirrors can attain scientific feasibility according to the definition given here, they should be the most likely reactor candidates. The questions are whether direct energy conversion can be developed at a reasonable price; whether the magnetic field is efficiently used (that is, cheap enough); and of course radiation damage.

The Astron seems heir to more difficulties: the size may be very large, and it is not at all clear whether relativistic-energy, high-current guns will be cheap enough. Direct conversion is still a problem.

Even if a continuous flow pinch, 10 m long, can be developed, it is doubtful that an economic fusion reactor can be made of it. The power density is immense, and presumably an exceedingly high magnetic field is needed to confine the plasma string. Could this ever be done without putting the field coils near the plasma, thus exacerbating heat transfer and tritium regeneration problems? There are more problems besides.

Several quite different schemes for achieving controlled fusion are not shown in Figure 9; the so-called "lasar ignition" scheme deserves mention (Holcomb, 1970). In that, the pulse from an ultra-high-energy laser is focused on a small pellet of solid D-T and heats it to fusion temperatures before the pellet has time to disassemble. The disassembly speed is about 10^6 m/sec at fusion temperatures, and the pellet size is the order of 1 mm. Thus the main heating pulse must be less than 10^{-9} second long. Even more, the most efficient heating scheme involves using several smaller preheating pulses to set up initial temperature and density gradients in the pellet, and these must be applied with temporal accuracy of perhaps 10^{-11} second. These requirements can be met. About 10^5 joules is the minimum estimated to be necessary for energetic breakeven; enough fusion energy out to equal the laser energy deposited. Even these large values are not discouraging; what seems to me very difficult is producing power cheaply enough; for reference, 5×10^7 joules of such "explosive" raw heat deposited in (say) lithium coolant is worth about \$0.01; can one do all this repetitively with an expensive and fragile device?

Many of the questions raised above will require systems research, systems development, and systems engineering to answer. These arts have been put secondary to plasma research and experimental device development up to now.

TIME SCALES

Present pressurized water or boiling water nuclear reactors are satisfactory as interim devices, but their relatively low thermal efficiency and inability to breed much nuclear fuel (from ^{238}U or thorium) condemn them to a brief existence in our society, unless much more uranium is found. The total installed capacity of such devices will be much less than that of fossil fuel plants, so complaints about them are and should be based on relatively local considerations—for example, thermal effects in Biscayne Bay. These words should in no way be taken as denigration of the validity of local complaints.

The view here is broader, and of longer time scales. The real question concerns second-generation fission breeder reactors (for example, a liquid metal fast breeder, or molten salt breeder) vis-a-vis the possibility of controlled fusion. At one time it was thought that fission suffered a relative disadvantage of insufficient nuclear fuel because of lack of uranium in the earth's crust, whereas deuterium is in plentiful supply. This is not true; there are adequate supplies of ^{238}U and ^{232}Th, D, or ^6Li for some 10^8 or more years of society based on high energy consumption. Even better, all these are resources for which little alternate use is forecast.

The real questions of fission breeders versus fusion breeders (which have to breed their tritium, as we have seen) involve feasibility, relative cost, time scales, and environmental factors, which all tend to be related. We have discussed the first of these topics and will not return to it in detail. To put the costs in some perspective, it should be pointed out that an additional penalty of \$20 per thermal kilowatt—that is, doubling the maximum cost mentioned earlier—would add by itself less than \$2 per month to the present average residential electric power bill. That is no invitation to adopt expensive options thoughtlessly—as electric power use increases, extra costs hurt more—but it *is* a way of saying that substantial changes could be afforded in reactor cores (fission or fusion) *if* even moderate social benefits were likely to accrue. That view will affect later remarks.

With regard to time scale, there is some real misunderstanding. Controlled fusion is *not* an alternative to the first-generation fission breeders, as was at one time thought. The question is whether fusion or some second-generation fission breeder will be preferable. The time scale goes like this: even if scientific feasibility is demonstrated by 1975, basic studies related to topics above the waterline in Figure 9 will occupy several years beyond. After that, at least one pilot-model fusion device would occupy our attention until the mid-1980s; then fission reactor experience shows that the lead time is long for designing and building the economic plants to follow. It is possible that fusion power will be available in appreciable quantity by the year 2000, even with a fortunate outcome along one of the paths in Figure 9. A few optimists propose 1990; pessimists propose never.

This long time before beneficial installation might seem to permit a comfortable period of grace before basic decisions about the overall feasibility and future of fusion need be taken. That is not so: other time scales enter. An important one is the fact that present gas diffusion plants for uranium enrichment may reach the end of their life by about 1990. First-generation fission breeders will have come into service well before then, but large, new, gas diffusion plants will still be needed. The question is in part whether the replacements are for an interim continuation, for a long-term continuation, or something else. Such expensive construction (several billion dollars) and the concommitant commitment bespeak a fairly clear decision by 1980 about what is to be built. For that, relative rank ordering of nuclear power systems will be needed several years earlier. Thus important decisions need to be made about the relative merits and eventual feasibility of nuclear power systems in the next few years. When the decisions start to be made, it becomes increasingly difficult to alter the course of events, because large economic and intellectual investments start to be made in the chosen course, and it usually is easier to stumble forward than to reach back. In truth, controlled fusion must from here on be subject to increasingly detailed technological assessment. To be late or unresponsive in this activity is to risk being irrelevant.

HAZARDS

Upon the topic of the next two sections, much arrant nonsense has been written, reminiscent of Ben Johnson's *The Alchemist.*

Almost everyone agrees that the most appreciable nuclear hazard of controlled fusion is that of tritium. A 5000-Mw (thermal) fusion plant would cycle about 10^8 curies of tritium through the plasma per day at 0.05 burnup, and actually burn 5×10^6 curies per day. How big will the inventory be? That depends on the rapidity with which unburned tritium can be reclaimed from the plasma pumps and the efficiency with which regenerated tritium can be scavenged from the moderating blanket. What little has been done on the pump problem suggests that something like one day's throughput may be held up in transit between exhaust from the fusion plasma and reinjection. For the blanket, more thoughtful analysis (Fraas, in Hall and Maple, 1970) suggests that 10 or 20 days of bred inventory may be held up in the huge bulk of lithium coolant, graphite, and so forth. At 0.05 fractional burnup, the two inventories would be about equal: a total of 2×10^8 curies.

This is a lot of radioactive material, comparable (in curies) to the amount of the most hazardous fission product ($\approx 10^8$ c of ^{131}I) expected to be found in a fission breeder reactor of the same size. But after that the comparison is not parallel. Per curie, tritium is relatively benign (9 kev average energy $\beta-$) and in the gaseous form is only weakly biologically active. Then to this stage in the discussion, the relative hazards of fusion versus fission are perhaps $1 : 10^5$; on that basis fusion reactors could be installed anywhere without any containment shells (Fraas, in Hall and Maple, 1970). Still extreme care must be exercised.

Complicating this story are the starting-to-be-assessed hazards of tritium being released as T_2O, or tritium leaking through the reactor structure, and the like (Morley and Kennedy, in Hall and Maple, 1970; Fraas and Postma, 1970). For the first, T_2O enters the life cycle as does water, which increases the relative hazard considerably. For the second, hydrogen (hence tritium) delights in diffusing into and through metals, much more so than does any other element. This is no hazard of critical nuclear accident, but rather the problem of preventing the plant from having radioactive B.O. It can be solved technologically, for example, by placing vacuum barriers at critical places where tritium will migrate. But what will it cost? For example, if the fusion system cost including all such protective arrangements equals the cost of a liquid metal fast breeder plus a carefully prepared hole beneath the city to hold it, any advertised safety advantages of nuclear fusion become hard to see.

These tritium migration and scavenging problems are now starting to receive some attention, and in a few years a lot more can be said.

In the meantime, it is possible that fusion will retain a substantial advantage, which will be reflected in a price differential of $10 to $20 per thermal kilowatt.

Another nuclear nuisance is that the 14-Mev fusion neutrons will make the basic structure of a fusion reactor highly radioactive. Fission reactors have the same problem; the components are in no danger of being spread through the environment, so this activation poses more of a maintenance problem than a hazard.

About nonnuclear accident hazards, fusion and fission seem to be a stand-off; one uses large amounts of liquid lithium or fused salts; one uses similar amounts of sodium. These hazards seem small, perhaps less than those enjoyed by people who live next to railroads on which many things are transported.

Permanent storage of long-life fission products is an additional problem for fission reactors; the advantage to fusion is modest, because total storage charges are expected not to be severe (on the scale of things discussed here).

OTHER ENVIRONMENTAL AND TECHNOLOGY ASSESSMENT QUESTIONS

Arguments about fossil as compared to nuclear power have often been made in terms of which kind of plant should be installed somewhere remote from population centers. As a corollary, the environment is imagined to be restored by having many nuclear power plants at remote locations producing electricity, which is transmitted to load centers.

That is all very well, but some kind of Sutton's law (After Willie Sutton who, on being asked why he robbed banks, replied, "That's where the money is.") suggests that we look at the heart of the problem, which is elsewhere. Most people in the United States and other developed countries live in cities. Predictions vary for the energy requirements in (say) 1980, but all agree that even with the trend toward electric power accounted for, the nonelectric energy requirement will exceed the electric energy requirement by nearly an order of magnitude. Much of this nonelectric demand is for transportation. But even space heating, industrial process heat, and so forth still add up to much more than the predicted electric demand, and all this is now supplied by fossil fuels. Therefore, if fossil fuels are to be substantially traded for nuclear ones, nuclear power plants must be built in or very close to population centers. The question of hazards and the cost of assuring safety discussed in the previous section must be looked at from this point of view.

Analysis of the total social costs and benefits is complicated enough for fission breeders versus fossil plants, and is yet in a primitive stage. Including fusion as an option will make further complications. Either advanced fission breeder reactors or fusion reactors are expected to have good thermal efficiency; some propose 50% or more (compared with about 32% for present reactors, 41% for present fossil fuel plants, perhaps 50% for advanced ones). Proponents of fission breeders promote that the total environmental difficulties and social cost of nuclear power are substantially less than those of fossil fuel plants. We agree with this when the various diseconomies—those charges put upon the public sector and not now made a charge on the generating company—are included. That is, the effects of sulfur and nitrogen oxides, and of particulate emissions, place considerable burdens upon us as a whole; the country is taking steps to deal with them, and the curative costs are very large.

Beyond that, many more factors enter; here are some. Strip mining of coal can despoil large tracts of land for long periods. Deep mining of coal or uranium is hazardous; lithium mining also brings problems. Any fission reactor located on the surface in a city probably must have an exclusion area around it. Analyses show that this valuable land can be used for some agricultural purposes, very possibly in combination with some of the reactor's waste heat (Milhursky, 1970). But even if no direct economic use of the land is made, what large city could not do with an internal area having a pleasant vista? It is hard to quantify such social values, but surely they are substantial; recall the view down the Serpentine from Kensington Palace in London. Plant size and tradeoffs between capital cost and fuel cost can and should have substantial leverage on proper urban planning, but so far they do not. For example, large plants with low fuel cost could afford to be run with a policy of very cheap (free to some users?) off-peak power. With such a policy, different activities and living prospects can be stimulated in cities. The well-known positive feedback—via larger plant size, hence lower unit electricity cost, hence increased demand and accelerated technology change toward electricity—involves assessing much of future technology: can transportation be based on some electric process, for instance?

Even fission and fusion are by no means mutually exclusive choices. They might complement each other, because fusion is predicted to have a large available neutron excess, and some otherwise attractive fission breeder schemes look dubious because the fuel doubling time is too long(Lidsky, in Hall and Maple, 1970). Can fusion reactors then be used to manufacture incremental fissionable material, hence bringing about a useful symbiosis?

Yet all this does not reach the deepest layers of the problem. If we assign importance to the fact that controlled fusion could supply our energy needs for aeons, we should also see what constitutes the energy policy. Just producing more is clearly inadequate; using it sometimes brings difficulties too, such as the summer temperature rise in ghetto streets because of operating air-conditioners. Then should we reduce energy dissipation by having better insulated buildings? Perhaps some principle of minimizing the entropy increase needs to be factored in. For fossil fuel utilization, this certainly seems required; jet plane travel is not wholly satisfactory, when almost as much fuel is burned per trip as if each and every passenger drove the distance by himself in his own automobile.

These are not empty phrases; if high speed intercity transport switches from aircraft to tunnel vehicles, substantial switch from fossil to nuclear (electric) power is possible. There is a lot at stake, an adequately broad assessment has not been made, and we are uncertain about what the policy ought to be. Indeed nowhere have problems on this scale—as they really exist in society—been approached in such an integrated fashion hitherto. This comment has broader implication than just to controlled fusion and relates to what appear to be very basic difficulties in how we organize ourselves to solve large societal problems. But that is another story (Rose et al., 1970).

It is in this broad context that controlled nuclear fusion will or will not be brought to fruition. We believe that, for fixed plant requirements, nuclear fission can be made substantially more attractive than can burning coal or oil, for most purposes. As implied in earlier sections, we also believe that the situation could be improved even more with successful fusion power. But these are still beliefs, not yet firm facts.

It would be rash to predict the outcome; not all schemes now being worked on will be adopted, which is the price in technology assessment of keeping options open. Surprises come, not all unpleasant, and a historic parallel occurs to me (Singer et al., 1954–58). In 1680 Christiaan Hüygens decided to control gunpowder for peaceful purposes, as a perpetual boon to mankind, and set his assistant Denys Papin to invent a controlled gunpowder engine. After ten years of difficulty, Papin had a different idea, wrote in his diary,

Since it is a property of water that a small quantity of it turned into vapour by heat has an elastic force like that of air, but upon cold supervening is again resolved into water, so that no trace of the said elastic force remains, I concluded that machines could be constructed wherein water, by the help of no very intense heat, and at little cost, could produce that perfect vacuum which could by no means be obtained by gunpowder.

then invented the expanding and condensing steam cycle, which made possible the industrial revolution.

REFERENCES

Alexander, T., 1970. *Fortune* 81, no. 6: 94–97, 126, 130–132.

Artsimovich, L. A., et al., 1969. *J. Exp. Theor. Phys. Lett.* 10: 82.

Artsimovich, L. A., et al, 1969. *Nuclear Fusion Special Supplement* 17.

Beach, A. D., et al. 1969. *Nuclear Fusion* 9: 215.

Beal, J. W., et al., 1969. In *Plasma physics and controlled nuclear fusion research conference*, IAEA, Vienna.

Bishop, A. S., 1958. *Project Sherwood, the U.S. program in controlled fusion.* Reading, Mass., Addison-Wesley.

Coensgen, F. H., et al., 1969. In *Plasma physics and controlled nuclear fusion research conference*, IAEA, Vienna.

Dandl, R. A., et al., 1969. In *Plasma physics and controlled nuclear fusion research conference*, IAEA, Vienna.

Fowler, T. K., 1969. *Nucl. Fusion* 9, no. 3.

Fowler, T. K., and R. F. Post, 1966. *Sci. Am.* 215, no. 6.

Fraas, A. P., 1968. *A diffusion process for removing tritium from the blanket of a thermonuclear reactor.* U.S. AEC Report on ORNL-TM-2358, Oak Ridge National Laboratory.

Fraas, A. P., and H. Postman, 1970. *Preliminary appraisal of the hazards problems of a D-T fusion reactor plant.* U.S. AEC Report ORNL-TM-2822, Oak Ridge National Laboratory.

Hall, J. L., and J. H. C. Maple, eds., 1970. *Nuclear fusion reactors.* Proceedings of the Energy Society Conference on Fusion Reactors, United Kingdom Atomic Energy Authority, Culham Laboratory, Sept. 17–19, 1969.

Holcomb, R., 1970. *Science,* 167: 1112.

Little, E. M., et al., 1969. In *Plasma physics and controlled nuclear fusion research.* Conference Proceedings IAEA, Vienna, vol. 2, p. 555.

Mather, J. W., et al., 1969. *Phys. Fluids* 12: 2343.

Mihursky, J. A., in preparation. *Summary of meeting on beneficial uses of waste heat.* Oak Ridge, Tenn., Apr. 20–21, 1970.

Post, R. F., chairman, et al., 1970. *Preliminary report of direct recovery study.* Report UCID-15650, Lawrence Radiation Laboratory, Livermore, Calif.

Rose, D. J., 1968. *On the fusion rejection problem.* United Kingdom Atomic Energy Authority Culham Laboratory Technology Division Report No. 82.

———, 1969. *Nucl. Fusion* 9: 183.

———, 1970. *The case for national environmental laboratories.* U.S. AEC Report ORNL-TM-2887, Oak Ridge National Laboratory.

Rose, D. J., and M. Clark, 1965. *Plasmas and controlled fusion.* 2nd revision. Cambridge, Mass.: MIT Press.

Rothman, M. A., et al., 1969. *Phys. Fluids* 12: 2211.

Singer, C. J., et al., 1954–58. *A history of technology,* vol. 4. London: Oxford Univ. Press.

Steiner, D., 1968. *Neutronics calculations and cost estimates for fusion reactor blanket assemblies.* U.S. AEC Report ORNL-TM-2360, Oak Ridge National Laboratory.

Taylor, J. B., 1962. *Nuclear fusion supplement,* part 2.

U.S. AEC, 1969. Report LA-DC-10618. Los Alamos Scientific Laboratory.

U.S. AEC, 1970. *Summary report: Use of steam electric power plants to provide low cost thermal energy to urban areas.* Report ORNL-HUD-14. Oak Ridge National Laboratory, in preparation.

Ware, A. A., and J. E. Faulkner, 1969. *Nucl. Fusion* 9: 353.

Yoshikawa, M., et al., 1969. *Phys. Fluids* 12: 1926.

MARC KRAMER, DAVID FENNER,
JOSEPH KLARMANN, AND ROBERT H. WILLIAMS

Geothermal Energy

One of the major concerns in the energy crisis is the fact that we are currently using energy resources at a rate that exceeds by many orders of magnitude that at which they are produced by natural processes. Two sources of energy exist which, in principle, can more than adequately supply our energy needs without being significantly depleted: the energy radiated by the sun, and the energy flowing in the form of heat from the center of the earth. Solar energy has the drawback that the energy density is low, so that economic exploitation appears difficult. In most cases, this is also true of geothermal energy, but there are some areas close to the surface of the earth which have such a high heat content that the geothermal energy from these areas can be economically exploited for electrical power production. The mechanism which causes these anomalies is still a matter of debate.

These areas, called geothermal zones, can be classified as either "dry" or containing natural ground water. Although "dry zones" are the most prevalent, exploitation of these zones appears rather difficult. The geothermal zones containing ground water are the main concern in this review. At present, only the most efficient of the reservoirs, those with high temperature steam (hyperthermal zones), are being exploited for electrical power generation, while lower temperature reservoirs are mainly used for heating purposes. However, research is being done on processes which would be capable of generating electricity from the low temperature zones. Since the energy can be utilized mainly near the source, the total amount of energy concentrated in the reservoirs, and the geographic distribution are important in assessing the significance of geothermal resources. Since geothermal exploration is still in its infancy, the estimates of the total energy available vary widely. Most

recent estimates indicate that geothermal energy resources from hyperthermal zones are commensurate with the worldwide hydroelectric
resources while the resources from low temperature zones are much
more extensive. Limited experience to date has shown that geothermal
power produced from high temperature steam fields is economically
competitive with fossil fuel power.

GEOLOGIC MECHANISMS

Description of Geothermal Zones

The normal temperature gradient[1] of the earth's crust is about 30°C/
km. This increase in temperature is widely believed to result from heat
generated by radioactivity in the earth's crust. However, in several
selected areas of the earth, there are abnormally high temperature
gradients reaching values as high as 300 to 500°C/km, where heat
flows towards the surface are 100 to 1000 times greater than normal.
These areas are referred to as hyperthermal zones. In another type of
geothermal area the temperature gradient is between 50 and 100°C/km
and temperatures are usually below the atmospheric boiling point of
water. These zones comprise a larger area than the hyperthermal zones
and are referred to as low temperature geothermal zones.

Most geothermal energy development has involved the hyperthermal
zone. However, very recently there has been discussion on how best
to tap the heat from both the low temperature geothermal reservoirs
and ultimately the normal heat gradient of the earth. Total potential
resources from the hyperthermal and low temperature geothermal reservoirs are vast and far outweigh any present day conventional energy
source.

Since the science of geothermics is rather new, our understanding
of the physics and geology of hyperthermal reservoirs is still evolving.
What we do know is that by some process (to be discussed later)
hot magma has lodged itself some 3 to 10 kilometers below the earth's
surface, thus providing the heat source for a geothermal reservoir.[2]
Overlying this magmatic body is faulted rock which is cracked enough
to allow water flow. Above this rock is an underground water body
commonly known as an aquifer, which is heated to the temperatures
of the surrounding rock. An impervious layer of rock caps the whole

---

[1]Temperature gradient is defined as the change in temperature per unit of depth.
For example, if at 100 meters below the surface the temperature is 30°C, at 200
meters 45°C, the temperature gradient would be .15°C/meter.
[2]Magma is the molten rock material within the earth from which igneous rock is
formed following cooling and crystallization.

reservoir. Geothermal energy is tapped by drilling deeply through the impervious layer into the aquifer zone. The reduction of pressure (from breaking through the cap) causes the high temperature water to flash into steam which can then be utilized for the production of electricity. There are, of course, variations to this description. For example, in the Imperial Valley, in California, the acquifer seems to extend to the magma zone. However, regardless of variation, a hyperthermal zone consists of a source of heat (magma), water, and an impenetrable cap zone that keeps the water under high pressure.

Another type of reservoir, which has received little attention among the geothermal experts, is the so-called dry geothermal zone, in which no ground water reaches the reservoir.[3] Until recently, these zones were considered useless since it was believed that any method of extracting the heat would be far too costly. Recent work in the field indicates that the costs may not be prohibitive, and several schemes for extracting the heat have been proposed. A more extensive discussion will be presented later.

Examples (Wairakei, New Zealand, and Larderello, Italy)

The following two examples illustrate, in more detail, the geology and structure of a hyperthermal zone. The Wairakei field in New Zealand is the second largest electrical producing hyperthermal area in the world and has been widely studied. The field is described in Figure 1.

---

[3]Unless modified by the word "dry" hyperthermal and low temperature geothermal zones refer to reservoirs which contain natural ground water.

Figure 1   Hyperthermal zone: Wairakei, New Zealand

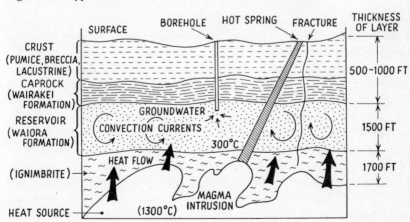

(a)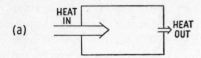

(b)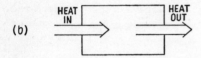

(c)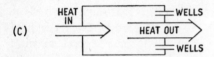

Figure 2   Heat flow diagrams for hyperthermal zones

Below the surface crust, a caprock formation of a relatively imper-
meable layer of rock acts like the top on a pressure cooker trapping the
geothermally heated water below. In New Zealand, this cover is thin,
cracks occur in it, and steam can be rising from the ground in a few
spots. Below the cap is a highly penetrable rock, quite permeable to
water flow, which forms the hot water layer containing the stored
energy which is utilized for electricity production. How the heat from
the magma reaches the aquifer zone is not entirely understood. If the
lower layer of rock were unfractured, the heat from the magma would
be only slowly conducted to the water above (rock is a poor heat conductor) and a geothermal zone would not exist. However, since the
lower layer is faulted, water from the lower part of the aquifer can
come into contact with either the magma or a thin rock layer surrounding the magma. Banwell (1962) has proposed several mechanisms by
which this heat transfer can occur.

The history of heat flow in a hyperthermal zone is indicated in Figure
2. The initial stage (Fig. 2a) probably occurred some several hundred
thousand years ago in the case of Wairakei, when heat flow to the
reservoir caused heating of the rocks and the water. At some point in
time (Fig. 2b) an equilibrium was established, wherein as much heat was
flowing into the reservoir as out. Most of the heat leaving the reservoir
appears in hot springs; in the Wairakei fields the natural flow is some
10,000 kilocalories per second (Banwell, 1962). When geothermal wells
were drilled, this equilibrium was disrupted; now more heat flows out
than into the reservoir (Fig. 2c). The amount of stored heat flowing out

of the geothermal wells in Wairakei is on the order of 5 to 10 times the normal heat flow.

To determine whether a reservoir is worth developing and at what rate it should be developed, it is necessary to estimate the amount of energy stored in the reservoir. For Wairakei, Banwell estimates that some 250 megawatts of electricity could be generated for the next 500 years. So far no lowering of the reservoir's temperature has been noticed, indicating that the reservoir is large. Also, since the water table is not now decreasing, ground water must be replacing the water which is being taken out of the wells for the electrical generating process.

A second example is that of the hyperthermal field in Larderello, Italy, which has been generating electricity for over 60 years (Fig. 3).

The field in Larderello differs from that in New Zealand not so much in structure, but in the thickness of the different layers. Instead of a thin upper layer, the aquifer is capped by a thick layer of rock. Consequently, little geothermal activity is noticeable at the surface. The water layer is thinner than that of Wairakei. According to Banwell, the depth of the magma in the Larderello field is over ten kilometers. One major problem with the Larderello field is that because of limited recharge permeability more water is being taken out of the reservoir than is flowing in. Consequently, the output of some wells has fallen and new wells are being drilled to maintain output. If reinjection of water back into the reservoir is not undertaken, the whole field may go dry.

An important distinction between Larderello and Wairakei is that Wairakei is a "wet steam field," while Larderello is a "dry steam field." In a wet steam field, the steam rising from the well is heavily mixed with hot water. This hot water must be separated from the steam before the steam can be used for electrical energy production. In the dry steam fields, the steam goes directly to the turbines with little alteration. The distinction between wet and dry steam fields is important because the

Figure 3   Hyperthermal zone: Larderello, Italy

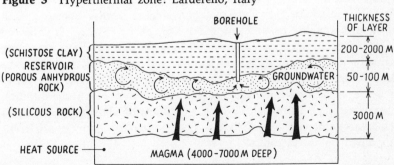

two types of fields have different economic opportunities and environmental consequences associated with them.

Sea Floor Spreading and the Creation of Geothermal Zones

Comprehensive new theories that rationalize large numbers of observations and explain major aspects of the physical world are rare in any field of investigation. Such a synthesis may be within reach in geophysics. The past few years have seen the emergence of a new theory concerning systematic movements of the sea floor. (Heirtzler, 1970)

Many geologists and geophysicists are beginning to believe that sea floor spreading or "new global tectonics" as some call it, is responsible for the existence of hyperthermal zones and many other geological phenomenon. While the concept of sea floor spreading is widely accepted, the particular effects of this process are still a matter of debate. Therefore, much of what follows in this section is still partly a matter of speculation.

The theory states, in its simplest form, that the sea floor is spreading from certain axes on the ocean floors. According to well documented evidence, this spreading occurs at a rate of from $\frac{1}{2}$ to 2 inches per year. This is quite a small distance over a single human lifespan, but over millions of years this spreading causes large changes in the surface patterns of the earth. The driving mechanism of this spreading is still a mystery, but, it is known that sea floor motion is a result of material rising from deep inside the earth along the oceanic ridges (labeled A in Fig. 4). This motion may be compared to that of a conveyor belt moving at extremely slow speeds. The new material arising from the earth pushes the older crust outward, perpendicular to the axis of spreading.

Along the oceanic trenches (labeled B in Fig. 4) the older oceanic crust is thrust back into the earth's interior. In essence, the oceanic ridges represent sources of new crustal material and the trenches represent sinks for the older material.

Most of the major hyperthermal zones have been associated with either land overriding oceanic ridges, or with oceanic trenches. The sea floor spreading hypothesis explains the distribution of these hot spots remarkably well. However, not all geothermal activity is directly associated with sea floor spreading. The low temperature geothermal zones seem to be associated with the geologic activity of mountain building (orogenic activity). Even in this case, some geologists are beginning to associate orogenic activity with sea floor spreading (now more commonly called "plate tectonics").

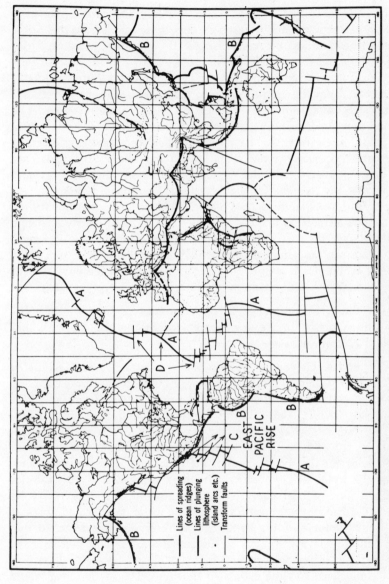

Figure 4 Worldwide location of sea-floor spreading (Reprinted from Stacey, Physics of the Earth, John Wiley & Sons, Inc., 1969, p. 196)

GEOTHERMAL ENERGY RESERVES

Estimates of Global Potential

Hypothermal Zones and a Basis for Estimates. The problem of estimating hyperthermal energy reserves closely resembles the historical problem of estimating oil reserves. As the petroleum industry developed, more and more oil fields were discovered and the reserve estimates sharply increased with time. In geothermics, resource estimates have increased one thousand-fold over the past decade. Only recently have investigations begun on the huge low temperature resources. The estimates presented below exclude the "dry hyperthermal fields" since these fields, until recently, have been considered useless as far as energy extraction is concerned. It is known that these "dry fields" are much more extensive than the hyperthermal zones containing ground water so that the reserves for the "dry fields" are much larger.

Reserve estimates are presented historically below. They exemplify the crude techniques used in making estimates and indicate clearly a change of attitude over the last decade. Hyperthermal reserves are no longer considered a rare and limited resource. Whether these reserves can be developed economically without significant damage to the environment is a question which will be discussed later in this report.

1961—At the UN Conference of New Sources of Energy, Ruiz Elizondo estimated that future development of geothermal energy would be around 3000 megawatts.

1965—In a U.S. Geological Survey report, Donald White estimated that 50,000 megawatts of electrical energy could be generated for the next 50 years from worldwide hyperthermal power development of known fields (White, 1965). Using the same analytical techniques employed by Banwell in estimating the reserves of Wairakei, White calculated the stored heat in all the geothermal reservoirs by taking an average value for the specific heat of the rocks and an average temperature for the wells. The stored heat in these wells is 2×10^{20} calories (30 billion tons of coal equivalent). Assuming that all the reservoirs of the world had not yet been found, White made an educated guess as to the actual volume of the hyperthermal reservoirs. He assumed that the known volume was 1/10th of the actual volume and therefore the global stored energy would amount to 2×10^{21} calories (300 billion tons of coal equivalent).[4] White estimated that only 1% of this heat would be tapped for geothermal power production. Removed over a period of 50 years, this heat energy could be used to generate electricity at a

---

[4]This is about 42 times the world's annual energy consumption.

rate of around 50,000 megawatts (assuming a power plant conversion efficiency of about 25%).

1966—C. J. Banwell, who has worked extensively with hyperthermal fields in New Zealand, attempted a calculation of geothermal reserves from a more theoretical approach (Banwell, 1966). He estimated that the amount of lava pouring out from volcanically active regions located in the Pacific coast zone has been about 393 cubic kilometers over the last 400 years. Taking a reasonable value of the heat capacity and the initial temperature of the lava, he concluded that the mean volcanic heat discharged over the last 400 years was 2×10^{10} calories per second. From a study of the heat discharged into hot springs and from other associated geothermal phenomena he estimated that about ten times more heat evolves from the cumulative buried magma in these areas than from volcanic activity. This is a conservative average since in some areas the ratio is much higher. Another important factor is that the potential output from a geothermal field is on the average about ten times that of the natural heat flux from the field because, when a geothermal field is drilled, much of the stored heat is released. Consequently, Banwell estimates that 2×10^{12} calories per second can be extracted in principle from all the hyperthermal zones of the Pacific.

Banwell was more optimistic than White and assumed that 10% of this heat could be recovered by commercial development. Consequently, his 1966 estimate for recoverable hyperthermal reserves in the Pacific region corresponds to an electrical power capacity of 250,000 mw which is about 20% of the 1970 installed electrical generating capacity of the world (assuming a power plant conversion efficiency of 25%). Banwell's Pacific region power estimate is thus four times larger than White's. However, Banwell estimated a depletion time of 200 years so that his recoverable energy estimate is actually 16 times higher.[5]

1970—More recent estimates seem to indicate even larger geothermal energy resources. Geologist Dr. Raymond Corcoran in a letter to the U. S. Committee on Interior and Insular Affairs has written, "It has been argued that geothermal power does not have enough potential to help solve the shortage of electricity the U. S. is expected to incur within a few years. Our studies indicate that the western U. S. could ultimately develop at least 100,000 Mw of electric power from geothermal steam, a figure that may not be sufficient to solve all our power needs but enough to be a full partner in the electrical power pool of the west." (U.S. Senate Subcommittee Hearings, 1970.) If the U. S. potential is

[5]Since the majority of hyperthermal zones are located in the Pacific, this figure can be used as Banwell's 1966 lower limit estimate on worldwide hyperthermal reserves.

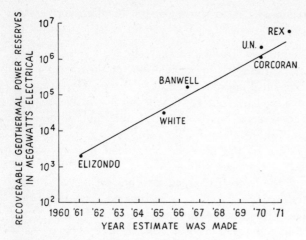

Figure 5 Hyperthermal reserve estimates

only 5–10% of the world's resources (as several geologists believe) then world reserves of the order of a million megawatts would be a conservative estimate.

1971—Robert Rex of the University of California at Riverside has recently presented even more optimistic figures for U.S. reserves. He stated that if intense exploration for hyperthermal resources were begun in the U.S., a million Mw of reserves could be located by the turn of the century (Rex, 1971). These reserves would imply a worldwide reserve figure on the order of ten million Mw.

Most current worldwide hyperthermal reserve estimates seem to be in the neighborhood of a million megawatts of recoverable power and many specialists in the field feel that one hundred thousand megawatts could be developed in the western United States. Presently, it appears that hyperthermal reserves are *at least* as large as hydroelectric reserves and as such can make a significant contribution to the satisfaction of electrical energy demand (Fig. 5).

Low Temperature Geothermal Zones and Normal Temperature Gradient Areas

The hyperthermal zones cover only a very small part of the land area. If we take account of the much larger areas of higher than normal temperature, which are estimated to cover some twelve million kilometers of land surface, the stored heat down to a depth of 3.5 kilometers is equivalent to a further 2×10^{13} tons of coal (40 million Mw over a 200 year period). (UN Memo, 1971.)

Little work has been done on examining either the wet or the dry low temperature geothermal zones. Considering the large quantities of heat which are generated during the geological process of mountain building enormous quantities of heat and energy are available. It is in mountain basins that most of these low temperature reservoirs are found. From 50–60% of the territory of the USSR is apparently underlain by thermal waters (low temperature geothermal zones) with an enormous total heat content. A large area in Hungary has been explored and over 60,000 square kilometers have been found to have an average thermal gradient of 50–58°C/km—nearly twice the normal gradient of the earth. Outside of Russia and Hungary, low temperature geothermal zones probably exist in areas of recent mountain formation such as the Himalayas in Asia, the Andes in South America and mountain ranges in Eastern Africa.

Most of these low temperature fields are "dry" and would require most costly and sophisticated engineering to extract the heat. Low temperature geothermal zones may in the long run prove more useful than the hyperthermal zones because they are more widely distributed and the reserves are larger. The development of these low temperature reserves for space heating has already proved useful in some parts of the world but the effort is still small compared to the potential. If economic and ecologically sound exploitation of these low temperature reserves can be developed, geothermal energy could become an important source of energy for the world community.

In the land areas of the world with a normal temperature gradient (30°C/km) the stored heat within the crust is extremely large, but hard to extract. According to C. J. Banwell, the French have begun to investigate a proposal to tap the heat of normal temperature gradient areas. "A pilot-scale heating project in the Paris Basin supplies 3000 flats with domestic hot water and part of their space heating from a large reserve at the depth of about 1800 meters. The water temperature is from 60–80°C, so the gradient is only slightly above normal." (Maugis, 1971.)

Though there has been some exploratory work in this area, exploiting the normal gradient of the earth for producing energy probably will not receive much attention in the near future. At present the cost of exploiting this resource is prohibitive in most cases.

World Distribution and Development of Geothermal Resources

A discussion of potential geothermal resources is meaningless unless the distribution of these resources is discussed, as unlike oil or coal, geothermal resources must be utilized close to where they are found.

Most hyperthermal zones are associated with the trenches and troughs involved in sea floor spreading and continental drift. The most active zones lie along the Pacific coast lines of South America, Central America, Japan, and the islands of the South Pacific. They run through Turkey, Italy, Iceland, parts of the Middle East and Eastern Africa. But most of the highly populated areas of the world do not have access to this resource. No hyperthermal activity has been found in the populated areas of India, China, of Asia, Europe, the eastern U.S., or in the eastern part of South America. Only in Japan, Central America, the South Pacific islands, and the western U.S. and East Africa, can large populations benefit from the enormous resources of these hyperthermal reservoirs.

Unless systems are developed which make electrical power transmission over long distances efficient and economical, the poor distribution of hyperthermal zones will render geothermal electric power from these zones important only in a few localized areas. This is another reason why more effort must be put into examining the more widely distributed low temperature zones.

Geothermal Energy Development in the United States

Geothermal energy development began in the early 1920s in California. However, it was rapidly abandoned because of a lack of sponsorship and technical sophistication. Until recently, geothermal energy was considered primarily a geophysical phenomenon of academic interest—not a likely major energy resource to be exploited for man's use. In the late 1950s, Magma Power Company was formed to develop geothermal power for commercial purposes. Steam recovered from drilling in the Geysers area (near San Francisco) was sold to Pacific Gas and Electric Company, which initially used it to run a small 12.5 Mw power plant. In 1961, Union Oil became interested in production and acquired a large tract of land in the Geysers area. Experience has proved the Geysers field to be one of the world's largest hyperthermal reservoirs. Large scale development is now underway in this field. Presently, Pacific Gas and Electric is purchasing enough steam to generate 82 Mw of electricity, and future purchases should run to over 1000 Mw. The known reserves of the field are estimated by some specialists to permit a generating capacity of 5000 Mw which is the output of this country's largest dam, Grand Coulee.

Another area in the west which has received widespread attention is that of the Imperial Valley in Southern California (near the Mexican border). Robert Rex and his associates at the University of California have done extensive studies on these fields and they have concluded that the reserves would permit 20,000–30,000 Mw of electrical power development while simultaneously producing from 5–7 million acre feet per year of desalinated water for irrigation purposes. This 1800 square mile reservoir is surely one of the biggest in the world, though some geothermal experts are skeptical of Rex's estimate of reserves. Full development of this field would be a large undertaking and some 2000–5000 wells would have to be drilled covering between 15–40 square miles.

A new geothermal development by the Magma Power Company and Union Oil is underway at the Brady hot springs in Nevada. Reservoir temperatures probably exceed 400°F. Magma Energy (a subsidiary of Magma Power Company) is planning to build an experimental unit in this area which would generate approximately 7.5 Mw of electricity.

Another area which shows considerable potential is the Casa Diablo region in Mono County, California. Both legal and environmental problems have prevented development there so far. The legal problem may have been resolved with the passage of the 1970 Geothermal Act. The environmental problems which concern the disposal of toxic wastes from the geothermal wells, could perhaps be remedied by direct re-injection of waste water back into the Casa Diablo reservoir.

Other areas of potential geothermal development are Beowawe, Nevada, Surprise Valley in Modoc County, California, Mono Lake Basin in Mono County, California, and areas adjacent to the Yellowstone National Park. Areas warranting further investigation (determined from geological evidence) are indicated in Figure 6 (Koenig, 1970).

Figure 6 Geothermal areas in the U.S.

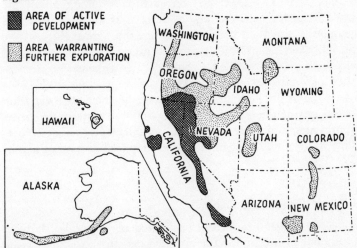

MECHANISMS OF POWER PRODUCTION

Exploration for Geothermal Areas

Various methods have been developed for geothermal exploration. They generally fall into the two categories of temperature-sensing methods and geologic structure investigations. The latter are similar to petroleum exploration methods, while the former involve chemical and gas content evaluation of water coming to the surface naturally, DC conduction and electro-magnetic methods used at the surface, and direct temperature measurement in boreholes (Bodvarsson, 1966).

The early stages of exploration may involve such methods as aerial infrared photography to map areas of higher than average temperature and surface heat flow measurements (Kiersch, 1964; White, 1969; Mount, 1969). The much more expensive temperature measurements in small boreholes have been shown to be the only really reliable methods of thermal mapping of a reservoir (Bodvarsson, 1970). The later stages of exploration require the drilling of a few wells of production size.

Geothermal Wells

The similarity between geothermal and petroleum wells is a considerable advantage to the geothermal industry, both technically and economically. Some refinements on drilling apparatus have been necessary to accommodate high temperatures and strong artesian pressures, but these refinements have been largely successful (Kaufman, 1964).

The greatest problem of geothermal wells is that of mineral precipitation in the well shaft, which causes an eventual choking off of the well. There is a tendency for minerals to dissolve at high temperatures in low lying regions and then precipitate when the temperature in the well drops as the water flashes into steam. In order to restore a clogged well to service, it must be redrilled at considerable expense.

The number of wells necessary to fully utilize the potential of a geothermal field tends to vary greatly because of the great variation in the structural characteristics of the field, the well's steam quantity, temperature, pressure, and lifetime. Thus it is difficult to predict beforehand the number, size, and cost of the wells required to exploit a geothermal field.

Geothermal Power Generating Plants

Geothermal electric power production originated in Larderello, Italy in 1904, as an offshoot of a boron extraction process. In subsequent years

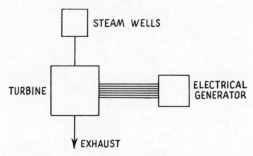

Figure 7 Non-condensing plant
(UN Conference, 1961, Paper #G/8 G/68/Kaufman)

the boron operation has given way to the increasing growth of geothermal power production.

Geothermal power plants are similar to fossil fuel and nuclear plants
in that they use high pressure steam to drive the plant's turbines. The
main difference is that geothermal plants derive their high pressure
steam from hyperthermal zones rather than from nuclear or fossil fuels.
In essence, the earth acts as a geothermal plant's boiler.

The plants in operation at existing geothermal sites are of two basic
types: noncondensing and condensing.

The noncondensing type (Fig. 7) is suitable only for small scale plants
in the range of 500 to 6000 kw. Its relatively low cost per kw of
capacity and its ability to operate on lower steam pressures makes the
noncondensing type of plant suitable for small test installations or for
use by the less-developed nations. It has a high rate of steam consumption and a low efficiency. In this type of plant, no mechanism for collecting the mineral content of the steam for disposal or exploitation is
feasible.

The condensing plant is similar to the noncondensing plant except
that the exhaust is condensed to water. The greater difference between
the input and output pressure and temperature of the turbine gives rise
to a substantially greater efficiency, so that fewer wells are needed per
unit of power output. The cost of the additional equipment for the
condensing plant can be offset by the greater power output of a given
well. The condensing type of plant can also be built on a much larger
scale, but this type requires a considerable amount of cooling water
and/or cooling towers (Fig. 8).

A variation on the condensing design that is used when the well
steam is very corrosive or has a high mineral content incorporates a
heat exchanger to produce pure steam from the heat of the well steam
(Fig. 9). The heat exchanger also provides a concentrated mineral

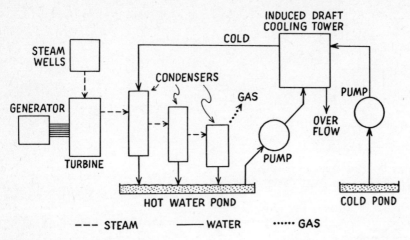

Figure 8   Condensing plant
(UN Conference, 1961/Kaufman)

impurity and gas exhaust to facilitate recovery of the minerals for sale.

Magmamax is a new process in geothermal energy development which may play an important role in the near future. In most major fields[6] the steam arising from the reservoir is mixed with superheated water (180–370°F) and only about 10–20% of this discharge, by weight, is steam. The Magmamax process would allow much greater energy recovery by extracting hot water rather than steam from the geothermal wells. This is done by keeping the pressure in the cycle high enough at all times so that the water never flashes into steam. The hot water is passed through a heat exchanger and a low boiling point fluid (freon and isobutane have been suggested) is heated and vaporized. The rapidly expanding vapors of the low boiling point fluid then turn the turbines. The process has many advantages, one of which is that if the water is reinjected into the ground the system operates in a closed cycle so that no pollutants enter the atmosphere.

Alternative Uses of Geothermal Energy

Even though a large portion of geothermal energy is presently being used for greenhouse and space heating, electrical energy production has been the dominant concern in the development of geothermal energy. Very little discussion was devoted to alternative uses of geothermal

---

[6]Of the many geothermal fields which have been found so far, only five produce "dry steam," unmixed with hot water.

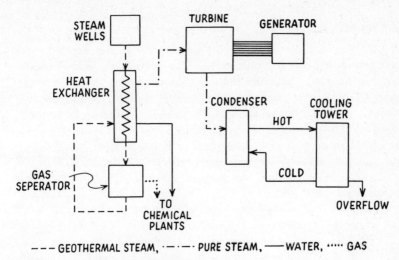

Figure 9   Condensing-heat exchanger plant
(UN Conference, 1961/Kaufman)

energy in the 1961 UN Conference on New Sources of Energy; however, by the time of the 1970 UN Conference in Pisa, the situation
had changed. The discovery of vast low temperature reserves has led to
an increasingly active interest in alternative uses.

The current aim of progressive planners in the geothermal area is to
utilize as much of the energy from the reservoirs as possible. This has
led to the concept of multidimensional development of geothermal
reservoirs. In particular, Joseph Barnea describes a UN project in Chile
which is being conducted along these lines (Barnea, 1972). The hot
water and steam from a wet geothermal field are separated. The steam
is used to produce electricity, while the hot water is evaporated to produce fresh water and minerals are recovered from the residue.

GEOTHERMAL-ELECTRIC POWER COSTS

The economics of geothermal-electric power production are still rather
uncertain. Numerous examples of operating geothermal plants are available, but the variation in the methods used to estimate costs is so great
as to make the estimates all but incomparable. A standardized method
of cost analysis would greatly assist in evaluating the economic feasibility of geothermal power (UN Symp., 1970). As yet no such system
has been put into use. However, the estimates made so far all indicate
that geothermal power is strongly competitive in cost with conventional
sources of electrical energy.

Cost of Exploration

At this time, exploration methods are still largely in the experimental stage and the costs generally unreported. The New Zealand geothermal team reported in 1961 that over the previous ten years, they had spent just under $1 million on exploration excluding the cost of drilling test wells. Joseph Barnea, head of the UN Resources and Transport Section, estimates the U.S. will have to spend from $1 to $2 billion to explore its geothermal resources. Few other estimates are available, indicating the uncertainty and variability of these costs. It is expected that in the future, when optimum exploration methods have come into wide usage, exploration costs may drop.

Well Drilling Costs

The costs of well drilling reflect the variability of the geothermal fields being developed (the depth and width of the well, the hardness of the rock to be penetrated). In 1961, the average well drilling costs were reported to be between $20,000 and $140,000 each (United Nations Conference, 1961). Wells drilled recently in Mexico appear to cost about $100,000 each, while the costs estimated for the wells being considered for the Imperial Valley of California will cost about $250,000 each. At this rate, drilling the estimated 2000 to 5000 wells needed to fully develop this region will bring the total well cost to $500–$1,250 million assuming there are no failures or necessities for redrilling (Rex, 1970). The 1970 UN Symposium on geothermal energy reported well costs of $98 per meter to $172/m in Japan, and $120/m average in New Zealand.

The estimated depreciation time is one of the largest variables in determining the cost for well drilling. In the U.S., a well is assumed to be productive for 20 years, in Japan 10 years, and in Iceland 7 years. After that time it is necessary to redrill the well or drill a replacement well. Without question, drilling is the most expensive phase of geothermal power production.

Power Plant Costs

Depending on the type of power production process used, the capital costs for generating capacity can vary. In general, the condensing plants cost considerably more per kilowatt of capacity than the noncondensing types. The 1961 UN Conference on energy reported the following costs for plants:

Noncondensing:
Larderello, Italy—$65 per kw installed
Mexico—$53 per kw installed

Condensing:
Geysers, California—$152 per kw
Iceland—$182 per kw
New Zealand—$106 per kw

Again, hidden variables can affect the cost greatly, for example, the Geysers plant would have cost 26% more if a new generator had been used in place of the old converted one actually used.

In 1970, the UN Symposium reported that generating capacity costs ran $105/kw to $300/kw for condensing types and about $80/kw for noncondensing types.

Considering the possible importance of plants designed to run on relatively low inlet temperatures, it is interesting that a plant being considered for Brady, Nevada, using isobutane as the working fluid at temperatures of 135°–200°C, is expected to cost from $120 to $160/ kw. Hence this type of plant may also prove to be economically competitive with conventional steam electrical power plants.

Other Costs

The cost of the pipeline needed to carry the steam from the wellheads to the power plant has been small at existing geothermal sites. Also, operation and maintenance costs have been low. Geothermal plants lend themselves to automation due to the small amount of operational supervision necessary. Maintenance requirements are generally low unless the steam is especially corrosive. Neglect of well repair can cause a blowout destroying the equipment, or necessitate periodic redrilling and cleaning.

The cost of land is rarely mentioned in economic reports on geothermal power. In the U.S., much of the land overlying known geothermal reservoirs is government owned. Consequently, Senator Alan Bible (D.-Nev.) has repeatedly introduced a bill since 1962 to authorize the leasing of land for geothermal exploration. In 1970, his Geothermal Steam Act was passed, providing for leasing of public lands to the highest bidder with a royalty on the steam removed plus a rental fee of at least $1 an acre per year until production wells are drilled (U.S. Cong. Rec., Sept. 1970).

At the time of the 1961 UN Conference, it was generally felt that the recovery of chemicals from geothermal steam was economically feasible at only a few exceptional locations. However, at the 1971 UN Symposium, considerable emphasis was placed on multipurpose plants and chemical recovery was discussed as a very likely economic asset.

In lower California where fresh water is scarce, the heat of geothermal fluids could be used to produce much needed pure water by distillation. The scale of fresh water could make the combined geothermal power and water plant very favorable economically (Rex, 1970).

Total Cost Estimates of Geothermally Produced Electricity

Total power generating costs are shown in Table 1 for three of the areas with the most experience at producing geothermal power (Kiersch, 1964). These figures are not strictly comparable, since, as mentioned above, a standardized method of cost analysis has not been put into use. Nevertheless, all of the available estimates appear to indicate not only that wet steam fields are economically competitive with conventional plants, but that dry steam fields are significantly less expensive (Tables 1 and 2).

Table 1[a]   Generating Cost in Mills per kw Hour

Larderello, Italy		
Noncondensing 4–6 Mw	2.74	
Noncondensing 16 Mw	2.60	
Condensing 128 Mw	2.38	Dry Steam
Cond. & Heat Exchange 84 Mw	2.96	
Wairakei, New Zealand		
Condensing 192 Mw	4.45	
Condensing 282 Mw	4.12	Wet Steam
The Geysers, California		
Condensing 28 Mw	2.36	Dry Steam

Table 2[b]   Economics of Geothermal Power

Energy Source	Mills/KWH (1)
Geothermal (Dry Steam)	2–3
Conventional Thermoelectric	5.47–7.75
Nuclear	5.42–11.56
Hydroelectric	5–11.36
(1) 92 Per Cent Plant Factor	

[a]Facca; Ten Dam, 1965.
[b]Whiting, 1968.

ENVIRONMENTAL PROBLEMS

Geothermal energy has been portrayed by some as having almost no adverse environmental effects. Though geothermal energy may have some important environmental advantages over our conventional power systems, there are still serious problems which must be carefully considered.

All geothermal steam contains some quantity of minerals (brine) as well as noncondensable gases such as: H_2S, CH_4, and NH_3. While current usage has not proven to be detrimental to the environment (mainly because of the small scale of operations) there is a potential for environmental impact when large scale operations begin. The environmental impact statement accompanying a proposed power plant should therefore list the nature and amount of contaminants expected and the method of disposal along with sufficient information to insure that the planned disposal will have only limited impact on the environment.

A most serious problem confronting geothermal development is the disposal of the geothermal waste water (water unused or condensed from the steam after passing through the generating turbines). In many cases, this waste water contains boron and other chemicals which would be harmful if disposed of directly into the surface water. In "wet steam fields," the mineral content is higher than in "dry steam fields," and in some areas the salinity of the surplus water approaches that of the ocean.[7] In these cases, direct disposal of the geothermal waste water into surface waters could have very serious consequences. One solution has been to reinject the surplus water back into the geothermal reservoir. However, great care must be taken in order that this waste water does not mix with the underground water which is needed for agricultural and home use. Reinjection has been started in El Salvador and in the Geysers area with initial success (Barnea, 1971).

In extracting great quantities of water from underneath the land there is a possibility that a general subsidence would occur. This possibility must be examined anew for each reservoir. Reinjection of the waste water back into the reservoir may minimize this problem. It is also possible that withdrawal or reinjection of fluid will cause earthquakes in a similar manner that brine injection has caused in the Los Angeles area (Hamilton and Meeham, 1971).

---

[7]In the Salton Sea area (near the border between Mexico and California), the salinity is so high that geothermal development is impossible. One of the reasons why the reserve estimates of Robert Rex for the Imperial Valley are regarded as too high is that his critics believe that the salinity may be too high for commercial development there.

Geothermal energy recovery requires extensive land use. Rex is considering development of some 15–40 square miles for maximum development of the Imperial Valley. There is no doubt that such development would affect wildlife and alter existing settlement patterns. "Commercial and residential usage would be generally incompatible with a geothermal field" (Goldsmith, 1971).

The thermal pollution from the waste heat is similar to that of other power plants. Cooling towers could provide a partial solution.

Finally, if a noncondensing plant is planned, the noise of the release of steam could become a serious annoyance to nearby residents. (See Papers GR/4 and G/18 in Conf., 1961.)

The environmental effects of geothermal power appear to be less severe than those associated with fossil fuel and nuclear power plants. However, before large scale geothermal development is undertaken, a thorough environmental impact study of geothermal energy should be undertaken.

DRY GEOTHERMAL ZONES AND NUCLEAR STIMULATION

Most hyperthermal and low-temperature geothermal zones are not accessible to ground-water circulation, and so yield neither steam nor hot water. At least in principle, the energy in these dry geothermal reservoirs can be extracted if somehow cold water can be fed into them, circulated, and withdrawn either as steam or as hot water. The principal difficulty here is that since the thermal conductivities of rocks are typically very low, the extraction of heat from rock at a usefully high rate therefore requires a very large heat-transfer surface. To produce the required surface area by conventional drilling or tunneling methods would be so difficult and expensive that it has never been seriously attempted. Recently, however, two unconventional methods have been suggested, one of which would use a nuclear explosive to create a rubble-filled cavity in the hot rock, while the other would use hydraulic pressure to produce a very large fracture system through which water would be circulated.

The suggestion has been made that one or more nuclear explosives be detonated 1-2 miles deep in a "dry" geothermally active region that does not naturally have porous rock. The explosion would create a highly porous region into which water would be injected from the surface. The naturally occurring heat would warm the water. The hot water and steam would be drawn off as with a natural geothermal system and used to generate electrical power, whereupon it would be reinjected.

This method of obtaining geothermal power raises several serious problems. Despite all the precautions taken, underground nuclear bomb tests in the past have often released some surface radiation. Since geologically active areas and geothermal zones invariably coincide, radiation from the explosion and debris may contaminate ground water in the area through fissures (natural or bomb induced). It has been the experience in geothermal well drilling that blowouts (ruptures of the well caused by high pressures) cause rather complete destruction of the drilling rig and necessitate further drilling before capping is possible (see papers G/43 and G/51, UN Conf., 1961). The consequences of such an occurrence with radioactive debris being vented into the atmosphere would indeed be disastrous. Finally, the most recent AEC study of nuclear stimulation for geothermal power showed costs too high to be competitive with fossil fuel power plants (AEC, 1971).

A second possible method of producing the heat-transfer surface required to exploit a dry geothermal reservoir is now under preliminary investigation at Los Alamos Scientific Laboratory. This method does not use explosives of any kind, but instead uses hydraulic fracturing to produce a very large crack system that connects two holes of unequal depth drilled into the reservoir. Fracturing is accomplished by using a high-pressure pump at the surface to produce high hydrostatic pressure in a sealed-off section near the bottom of the deeper hole. Experience in the oil-fields (where this is a common method of well-stimulation) shows that when the fluid pressure in the hole exceeds the sum of the horizontal component of the overburden pressure and the tensile strength of the rock, cracking begins at the borehole tip. Even in very deep holes, the "breakdown" pressure is of the order of only 10,000 psi above the hydrostatic head of water, and once formed the cracks can be extended to very large diameters by continued pumping at reduced pressure. The circulation loop would be completed by a primary heat-exchanger at the surface. Because of the heat-transfer advantages of water over steam, a closed, pressurized water system would be used. Once flow had been established through the underground system, the density difference between cold water entering the deeper hole and hot water rising through the shallower one would maintain convective circulation, and no pumping would be required.

A successful system of this type would, for example, produce about 260 Mw of thermal energy for approximately ten years from a single pair of 12-inch diameter holes and a hydraulically fractured crack with a radius of 4000 feet. However, it is anticipated that volume contraction of the reservoir rock as it is cooled will cause thermal-stress cracking to occur, which will extend the circulation system outward and downward—into regions of continuously increasing temperature. If this occurs, the useful lifetime of the above system will be very much longer than ten years, and its quality may actually improve as energy is withdrawn from it. The feasibility, efficiency, and economics of this method of developing dry geothermal reservoirs have yet to be demonstrated. It appears, however, to offer the possibility of extracting geothermal energy anywhere in the world that hot rock exists at a depth that can be reached by modern drilling methods (*The Atom*, 1971).

Though the "dry field" reserves are large, it is unlikely that geothermal energy development will be expanded into this area until more work is done on the hyperthermal and low temperature reservoirs which contain ground water. If geothermal energy becomes a large enterprise and geothermal technology improves, then exploitation of these "dry fields" may become a reality.

DEVELOPMENT OF GEOTHERMAL POWER

Geothermal power is a relatively cheap and abundant source of energy. The generation of power from established sources of geothermal steam poses no serious problems, but on the other hand, exploration for resources, and the development of methods for utilizing low temperature fields need stimulation. It is clear that in this country the electric power companies do not have the resources to carry out development of this nature. The oil companies have the knowledge needed for exploration but are not likely to encourage competition for their main product. Since full utilization of geothermal resources is an attractive alternative to the rapid depletion of fossil fuel resources and to the generation of vast amounts of long lived radioisotopes in nuclear reactors, active stimulation of development by investment of capital should be strongly encouraged.

REFERENCES

The Atom, 1971. vol. 8, no. 10, December.

Banwell, C. J., 1962. Thermal energy from the earth's crust. *New Zealand J. Geol. Geophys.* 6: 52–62.

———, 1963. Thermal energy from the earth's crust. *New Zealand J. Geol. Geophys.* 1, no. 3.

———, 1966. Geothermal power. *Impact Sci. Soc.* 17, no. 2: 151.

———, The efficient extraction of energy from heated rock, Part 2, (unpublished).

Barnea, Joseph, 1972. Geothermal power, *Sci. Am.* 70 (Jan.).

Bodvarsson, G., 1966. *The ore bin* 28, no. 7: 117–124.

———, 1970. *Geoexploration* 8: 7–17.

Bowen, R. G., and E. A. Groh, 1971. *Technol. Rev.*, Oct./Nov., p. 42.

Bradbury, J. J. C., 1971. The economics of geothermal power. *Nat. Resources Forum*, United Nations vol. 1, no. 1.

Carlson, R. H., 1959. Utilizing nuclear explosives in the construction of geothermal power plants. U.S.A.E.C., 2nd Plowshare Symposium, San Francisco, 1959 Proceedings, Part 3, pp. 78–87.

Elder, J., 1966. Heat and mass transfer in the earth: Hydrothermal systems, New Zealand.

Elders, W., et al., n.d. Crustal spreading in Southern California (the Imperial Valley is a product of oceanic spreading centers acting in a continental plate). The Institute of Geophysics and Planetary Physics, Riverside, California.

Engineering News Review, 1971. Geothermal resources gather a head of steam, May 6, pp. 30–36.

Facca, G., 1970. General report on the status of world geothermal development. Section 2, United Nations Symposium on the Development and Utilization of Geothermal Resources, Pisa, Italy, 1970.

Facca, G., and A. Ten Dam, 1965. *Geothermal power economics.* U.S. Senate Hearings before the Subcommittee on Minerals, Materials, and Insular Affairs, 89th Congress, 1st Session (on S. 1674), July 22, 1965, Appendix.

Godwin, L. H., et al., 1971. U. S. Geological Survey Circular no. 647.

Goldsmith, M., 1971. Geothermal resources in California: Potential and problems. Environmental Quality Laboratory, Report 5.

Hamilton, D. H., and R. L. Meeham, 1971, *Science* 172, no. 3981 (Apr.): 33.

Heirtzler, J. R., 1970. Sea floor spreading. *Sci. Am.* 219: 60.

Kaufman, A., 1964. Bureau of Mines Information Circular no. 8230.

Kiersch, G. A., 1964. Geothermal steam. Air Force Study, Cornell Univ., 1970.

Koenig, J., 1970. Geothermal exploration in the western United States. U.N. Geothermal Symposium, Pisa, Italy, 1970.

Lear, J., 1970. *Saturday Review*, Dec. 5: 53–61.

Lessing, Lawrence, 1969. Power from the earth's own heat. *Fortune 79.*

McNitt, J., 1969. Outlook for geothermal energy, Future Energy Outlook, The Colorado School of Mines.

Maugis, P., 1971. *Ann. Mines* May: pp. 135–142.

Meidev, T., and Robert W. Rex, 1970. Investigations of geothermal resources in the Imperial Valley. Univ. of California at Riverside, July.

Mount, R. L., 1969. Science forum. *Canada* 2 No. 2: 20–22.

Oil and Gas Journal, 1964. Soviets eye geothermal development. 62, no. 30 (July): 122–124.

Rex, Robert W., 1970. Investigations of geothermal resources in the Imperial Valley. Univ. of California at Riverside, June.

———, 1971. *Bull. At. Sci.* Oct.: 52–56.

Scott, Stanley, Se. E. Wood, 1972. California's bright geothermal future. *Cry California*, Winter: 10–23.

United Nations Conference on New Sources of Energy, vols. 2 and 3 (Geothermal Energy), 1961. Rome, Italy.

United Nations Symposium on the Development and Utilization of Geothermal Resources, (Summary Volume), Pisa, Italy, 1970.

United Nations. Memo, unpublished, 1971.

U.S.A.E.C., 1971. A feasibility study of a plowshare geothermal plant, April. A joint effort of: American Oil Shale Co., Battelle-Northwest, Westinghouse, U.S.A.E.C., Lawrence Radiation Labs, and Nevada Operations Office.

U. S. Senate, 1970. Hearings before the Subcommittee on Minerals, Materials, and Fuels of the Committee on Interior and Insular Affairs, 91st Congress, 2nd Session on S. 368, July 17–18.

U. S. Congressional Record—Senate, Sept. 16, 1970, S15758-S-15762.

White, D. E., 1965. Geologic Survey Circular no. 519.

———, 1969. *J. Geophys. Res.* 74, no. 22: 5191–5201.

Whiting, Robert L., 1968. Geothermal energy and resources. *Oil Gas Compact Bull.* 27, no. 1: 17.

Weismantel, G. E., 1971. *Chem. Eng.* Nov. 30: 24–25.

ARTHUR R. TAMPLIN

Solar Energy

This chapter discusses various schemes that have been proposed for the utilization of solar energy. The first section will discuss physical systems and the second section will treat biological systems. The major focus of the chapter will be to present a means of comparison; consequently the technical description will be somewhat brief. More detailed technical discussions can be found in the cited references.

In his June 4, 1971, energy message the President stated, "The sun offers an almost unlimited supply of energy if we can learn to use it economically." This statement reflects the increased interest in solar energy technology that has developed in response to the evolving energy crisis in the U.S.

Testifying before the Senate Interior Committee on June 7, 1972, Dr. Eggers of the National Science Foundation (NSF) stated:

"Solar energy is an essentially inexhaustible source potentially capable of meeting a significant portion of the nation's future energy needs with a minimum of adverse environmental consequences. . . . The indications are that solar energy is the most promising of the unconventional energy sources, and the foundation plans a substantial increase in fiscal 1973 funding of solar energy research to a total of 4 million dollars." (Eggers, 1972)

As an illustration of the potential of solar energy, consider that some 2 trillion kilowatt-hours (kwh) of electrical energy were consumed in the U.S. in 1970. Incident solar energy in U.S. deserts averages some 2000 kwh per year per square meter or 2 billion calories per year per square meter. (A calorie is the amount of heat needed to raise the temperature of one gram of water one degree centigrade.) In other words, our electrical energy consumption was equivalent to solar radiation falling on only some 400 square miles of desert. If this solar energy could be tapped with only 5% efficiency, just 8000 square miles of desert would be required (a 90-mile square). This is less than 10% of our deserts.

The three nonbiological classes of solar energy utilization are terrestrial, space, and marine. Terrestrial and space systems would use incident solar energy while marine systems would use both incident energy and solar energy stored in sea thermal gradients.

Essentially, two schemes have been proposed for terrestrial systems. One involves the use of solar cells and direct conversion of solar energy into electrical energy. The other involves the absorption of solar energy as heat which is either used directly or converted into some other energy form.

SOLAR CELLS

This technology received a substantial impetus from the space program and today, using silicon crystals, conversion efficiencies of 10% are routinely obtained. Systems using silicon solar cells have been proposed for electrical power generation (Glaser, 1971; Ralph, 1971; Cherry, 1971; NAS-NCR, 1972). At 10% efficiency, it would require only 4000 square miles of collector surface in the desert to generate the present electrical power consumed in the U.S.

This is essentially an on-the-shelf system. The major barrier to its use and, hence, the major area for research and development is in the fabrication of the cells. The cost of fabricating the silicon crystals is such that the overall system costs are prohibitive. Present costs for a nuclear plant are about $250 per kilowatt (kw) of installed capacity. (The $250 per kw costs are those estimated by the nuclear power industry, but probably are too low. At the same time, cost estimates made by proponents of solar power systems probably are also too low. As a result, the comparative basis for the relative costs of the two approaches may be adequate.) A silicon cell system would cost in the range of $100,000 per kw. Approaches for reducing this cost are discussed in the references above. It is suggested that it might be possible to bring the cost down to a competitive level (Glaser, 1971; Ralph, 1971; Arnold, 1972).

In addition to reducing the cost of fabricating the silicon cells, a savings could be achieved by using lenses to concentrate the sunlight and thus reduce the number of cells required. Moreover, the 10% efficiency is a factor of two to three below the theoretical efficiency of such cells. Costs could be reduced by improved conversion efficiency (Ralph, 1971; NAS-NCR, 1972). Finally, it may be possible to fabricate (at an economic cost) sandwiched cells which are able to utilize a greater fraction of the solar energy spectrum. One such cell could achieve an efficiency of 60% (Rappaport, 1972; Oak Ridge National Laboratory, 1972a).

Another possibility for reducing the cost of this system is the use of another type of solar cell (NAS-NCR, 1972). Considerable effort is now being expended on the cadmium sulfide cell (Oak Ridge National Laboratory, 1972b; *Energy Digest*, 1972a). The advantage of this material is that it functions as a thin polycrystalline film and hence does not require growth of large single crystals. As a result, the fabrication costs are expected to be at least 100-fold less than silicon cells. It is anticipated that these cells may become practical for individual homes (*Energy Digest*, 1972a).

Since these systems will produce electricity only when the sun is shining, they would either have to be augmented by other systems or would have to include an energy storage system. One storage system would involve the electrolysis of water and storage of hydrogen.

In summary, while solar cells represent an existing technology that has been used extensively in the space program, the high cost of cell fabrication is prohibitive for commercial power production. The cost of the cells would have to be reduced at least 100-fold. There are reasons to believe that this could be accomplished.

SOLAR HEAT SYSTEMS

These systems absorb solar energy as heat. The heat can be stored in high heat capacity materials in insulated containers and then subsequently used as heat or converted to electrical energy. Systems of this type fall into two classes: small systems for individual dwellings (solar home systems) and large commercial systems.

Solar Homes.　These systems simply absorb the solar energy as heat and store it in insulated bins as heated water or rocks. Such systems have been in use for some time for home water heating and even for space heating in homes (Proceedings, 1955; Tybout and Löf, 1970). They can also be used for air conditioning through the application of absorptive refrigeration (Löf, 1955). Space heating and air conditioning represent some 15% of our present energy consumption, electrical and otherwise (Eggers, 1972; Tybout and Löf, 1970). This is a substantial amount of energy; it is larger than our total electrical power generated today.

In some areas, these systems are already competitive with conventional systems (Tybout and Löf, 1970). As the cost of energy continues to increase and as solar technology is improved, these systems can be expected to come into wider use (Eggers, 1972; Proc., 1955; Tybout and Löf, 1970). It has also been suggested that a solar heat system could be coupled with a solar cell array and thereby supply a home with all of its power requirements (*Energy Digest*, 1972).

Commercial Systems.   A large commercial solar heat system has been proposed by the Drs. Meinel of the University of Arizona (Meinel and Meinel, 1972; Meinel, 1971). The system incorporates an advancement in solar absorption technology called "selective" surfaces. These surfaces have high absorptive properties but low emittances. Hence they would retard infrared re-emission as the temperature rises and thus produce a "super greenhouse" effect. Theoretically, such surfaces could be made to approach temperatures of 1000°F, but the present state-of-the-art falls below this, and some means of concentrating sunlight by a factor of two to four is needed. It is estimated that 90% of the incident solar radiation could thus be used. The energy would be stored in liquid sodium at a temperature of 1000°F. The heat would then be used to generate electricity through a steam turbine cycle. The Meinels estimate an overall efficiency of 30% for this system. An early estimate of the cost of this system was some $300 per kw of installed power as compared to $250 per kw for nuclear power (Meinel, 1971). Their estimate was based on using selective absorbers without concentrating lenses or mirrors. Nevertheless, they continue to propose that this scheme could produce competitive power when using lenses or mirrors (Meinel and Meinel, 1972).

The Meinels also present a plan for a million-megawatt average power (3.4 million megawatt peak power) national solar power system. This is equivalent to the projected electrical power production in the year 2000 at which time electricity is projected to supply 50% of our total energy requirements (USAEC, 1968–70; JCAE, 1971). At a hoped-for 30% efficiency, this system would require some 14,000 square miles (10% of our deserts). About one-third of this area would be occupied by the collectors. They also suggest that, as a by-product, enough fresh-water would be produced to supply 120 million people.

Another large-scale solar heat system has been proposed by Drs. Ford and Kane of the University of Massachusetts (Ford and Kane, 1971). The authors emphasize that this is only a proposal and thus uncertain. They suggest that sunlight could be concentrated by using inexpensive Fresnel lenses made of plastic. This concentrated energy could then be used to heat water to some 1500°C (2732°F), at which temperature a small fraction of water will dissociate into hydrogen and oxygen. The hydrogen would be absorbed into some chemical compound from which it could later be released. This system then would produce hydrogen as a fuel. They suggest that hydrogen could be marketed at an equivalent price with natural gas on a heat per pound basis. In this case they assume a 10% efficiency in the overall process.

To summarize, terrestrial solar heat systems offer the prospect of supplying, economically, a significant portion of this nation's and the world's energy supply. The 30% efficiency suggested by the Meinels would certainly greatly improve the prospects. At the same time, some combination of the Meinel and Ford-Kane proposals, vis-a-vis concentration by inexpensive Fresnel lenses, may lead to efficiencies greater than 10% and to a viable system. The ultimate economy of these systems would seem to depend upon developing the technology for producing economically competitive absorptive surfaces.

SPACE SOLAR CELL SYSTEM

A space solar cell system has been proposed by Dr. Glaser of Arthur D. Little, Incorporated (Glaser, 1968; 1971). This system would employ a number of satellites in synchronous orbits about the earth's equator. Solar cells would collect the solar energy and convert it to electricity. This energy would then be transmitted as microwave beams to collecting antenna arrays on earth. The receiving antennae would convert the microwave beams to direct-current electricity for transmission. This system would be a spinoff from the development of a space shuttle system with reusable boosters. With an existing space shuttle capability, it is estimated that this system would cost $500 to $1000 per kw of installed capacity. This cost estimate also depends on improvements in solar cell technology leading to higher efficiency and substantially lower fabrication costs.

MARINE THERMAL SYSTEMS

Two types of marine thermal systems have been proposed. One would utilize sea thermal gradients and extract the solar energy that is stored in the surface layer of the ocean. The other would use the oceans to support a floating platform and a system of concentrating mirrors to focus the incident solar energy on a boiler.

Sea Thermal Gradients.   A number of individuals have proposed the use of sea thermal gradients for the production of electrical energy (Eggers, 1972; Anderson, 1971; Heronemus, 1972). Most of these proposals parallel that of Anderson and Anderson (Anderson, 1971; Anderson and Anderson, 1966) which would use the temperature difference between the surface waters (about 79°F) and water at a depth of 2000 feet (about 43°F) in the Gulf Stream of Florida. The theoretical efficiency of a perfect Carnot cycle operating between this temperature difference is only 6%. Yet, as the authors state, since the primary concern is with economic efficiency and since there are no fuel costs, the system could be competitive. They estimate that such plants could be constructed for less than $200 per kw compared to $250 per kw for a nuclear plant.

This scheme envisages a nearly submerged platform floating in the Gulf Stream. The surface water is used to boil propane at high pressure. The propane gas is then used to power a turbine and generate electricity. The propane is subsequently cooled by the deep ocean water and returned to the boiler.

The use of ocean platforms would now appear to be a readily accepted concept. In addition to offshore drilling rigs and subsequent offshore oil field development, the establishment of ocean-based nuclear power plants is now a serious consideration. These sea thermal plants, therefore, are based upon extant technology for the most part. It would appear that only their precise economics are in doubt, but, even here, a doubling of the cost would not place them out of contention. Moreover, there are other potential features of these plants, which will be discusssed subsequently, that add to their overall prospects.

Sea Solar Heat Plant.   The system proposed by W. J. D. Escher is composed of a large free-floating ocean platform (Escher, 1971). Concentrating mirrors focus incident solar energy on a boiler system. This heat would then be used to drive a working fluid (water-steam) through a turbine generator. Again the cold water from the lower ocean layers would be used as a coolant. In other words, this is an ocean platform version of a terrestrial solar heat system.

The proposed output of this system would not be electricity. Rather, the electricity generated would be used to produce hydrogen by the electrolysis of water. The hydrogen would then be transported to use centers in cryogenic tankers. A rough cost estimate for this system is some $1500 per kw of installed capacity (*Industrial Research,* 1972).

The electrolysis of water to produce hydrogen as a fuel was also suggested as a possibility for the sea thermal gradient system (Heronemus, 1972). Hydrogen has the possibility of being distributed via pipelines at substantially lower cost than the transmission of electrical power. Moreover, it offers versatility as a fuel. It is ideal for fuel cell applications and in direct heat application; its only combustion product is environmentally pleasing water. Interest in hydrogen as a fuel is rapidly increasing (Jones, 1971; 1972).

Another feature of these ocean-based plants that adds to their potential value is the possibility of producing freshwater and of extracting minerals from the sea. Finally, it is suggested that these plants, by using deep waters for cooling, will produce an area of artificial upwelling of the nutrient-rich water. This, it is proposed, will enhance the productivity of marine life in the area. While it is possible that this will be beneficial, it is not guaranteed. Serious study of this potential ecological impact would certainly be necessary.

SUMMARY OF PHYSICAL SYSTEMS

There seems to be little doubt that the physical systems discussed above could provide a significant fraction of the world's energy needs in the future. Using a fraction of the desert areas in the Middle East, these oil-exporting nations could use solar energy to produce electricity and then electrolyze water. As a result they could export more BTUs in the form of hydrogen than they are presently shipping as oil.

The sea thermal gradient system is essentially at the engineering design and construction phase. The other systems require some research and development. The fabrication costs of solar cell arrays have to be substantially reduced to make them economically competitive. The commercial solar heat systems require the development of inexpensive selective surface collectors and lens or mirror concentrating mechanisms. None of this seems far beyond today's state-of-the-art, and the research and development cost would be minimal compared to most of today's programs. Most of these are already receiving funding from the National Science Foundation, the National Aeronautics and Space Administration, the Environmental Protection Agency (EPA), the Department of Defense, and state agencies. This funding involves improved solar cells, solar heat systems and the sea thermal gradient system (Oak Ridge National Laboratory, 1972b; Eggers, 1972).

BIOLOGICAL SYSTEMS

Not only our life, but also our way of life depends almost entirely upon the photosynthetic processes of plants. Plants, of course, have supplied us with food and an oxygen-rich atmosphere.

Moreover, our present stage of industrial development derives the great bulk of its energy from fossil fuels whose origin stems from the photosynthetic activity of previous ages. It is, therefore, not strikingly novel to look toward plants as a possible source of energy in the future.

However, whether or not this age-old friend of man will be given a share of our future energy markets will depend (and this seems somewhat profane) upon its economic viability in competition with the prosaic physical systems devised by man. In the preceding section, cost estimates per kilowatt of installed capacity were given for several of the physical systems for comparison with the costs of a nuclear plant. Note that these estimates excluded external or environmental cost wherein solar systems have a sizeable advantage. It is on the environmental cost side of the ledger that biological systems offer the most promise as energy sources.

ALGAL SYSTEMS

The efficiency with which plants convert solar energy into stored chemical energy will be discussed in a subsequent section where it will be shown that they are not competitive with the physical systems. However, a succeeding section will show that biological systems offer the greatest promise when deriving energy through waste management.

Photosynthetic Conversion Efficiency.   At the outset, it is useful to consider the photosynthetic efficiency of plants. Although the conversion efficiency is not an all-determining factor, it does relate to the size of the power system and hence to land use and capital cost. The physical solar systems are projected to have efficiencies in excess of 10%.

A simplified formula for the conversion of solar energy into stored chemical energy is $CO_2 + H_2O = CH_2O + O$ ($\triangle F^\circ = +114$ kcal).

When the caloric value of harvested crops is divided by the calories in the incident solar energy (about 1 billion calories per year per square meter) a conversion efficiency of only a few tenths of 1% is obtained from the normal practice of intensive agriculture (Meyer and Anderson, 1952; Kleiber, 1961a; Gaffron, 1960). Sugar cane is an exception and it has been produced with a conversion efficiency of a few percent. In fact, sugar cane wastes produce most of the electrical energy for the Philippine Islands (*Energy Digest*, 1972b). These efficiencies are, of course, influenced by the length of the growing season and by the growth, maturation, and death of the plant. At the peak of the growing season, considerably larger efficiencies are experienced (Wassink, 1953).

Theoretical or Maximum Efficiency.   Chlorophyll utilizes only that portion of the solar energy which has a wavelength between 400 and 700 millimicrons (Levine, 1969). This comprises only about 50% of the incident energy and sets an absolute maximum on the conversion efficiency. Moreover, it requires more than 114 kilogram calories (a kilogram calorie is the amount of heat needed to raise one kilogram of water one degree centigrade) of light energy to evolve 1 mole of O_2 by the equation given above.

Mole equivalents of blue light (450 millimicrons) and red light (700 millimicrons) are 64 and 41 kilogram calories, respectively. Experimental data and theoretical consideration (Gaffron, 1960; Levine, 1969; Seligen and McElroy, 1965) indicate that some nine quanta are required for each molecule of O_2 evolved. The conversion efficiency using red light is thus

$$\frac{114}{9(41)} \, (100) = 31\%.$$

This, then, leads to a maximum efficiency for conversion of the total incident solar energy of some 15%. This value is no better than that obtainable from physical systems.

Efficiency of Algal Systems   A considerable amount of experimental effort has been applied to the culturing of single-celled plants such as chlorella. The nutrient requirements for these algae are supplied by quite simple inorganic materials such as metal salts, phosphate carbon dioxide, and ammonia as a source of nitrogen. In indoor laboratory cultures using artificial light, conversion efficiencies of 15 to 25% are routinely achieved (Wassink, 1953; Tamiya, 1957; Meyers, 1958). The efficiencies measured in outdoor ponds vary from 2 to 10% for 400 to 700 millimicrons incident energy or 1 to 5% for total solar energy (Wassink, 1953; Tamiya, 1957; Oswald and Golueke, 1960; Kok and Van Oorschot, 1954). The lower efficiency of the outdoor ponds is mainly the result of variations in light intensity. In controlled experiments, the conversion efficiency increases with light intensity up to maximum and then declines (Wassink, 1953; Tamiya, 1957; Oswald and Golueke, 1960). The optimum conditions for conversion efficiency occur for only limited periods in outdoor ponds and hence, the conversion efficiency is lower.

Thus, algal systems are restricted to conversion efficiencies below 10%, probably closer to 5%. On the other hand, it is possible that as a result of future research, cell-free extracts can be prepared and stabilized which will carry out that portion of the photosynthetic cycle that results in the photolysis (chemical decomposition due to the action of light) of water. In this case, higher efficiencies may be obtained by the production of hydrogen as a fuel gas (Eggers, 1972). It is also possible to develop solar cells using such enzymes (*New Scientist*, 1972). Research in this area is being sponsored by the National Science Foundation-Research Applied to National Needs (NSF-RANN) and other federal agencies (Eggers, 1972; Oak Ridge National Laboratory, 1972b; *Chemical Week*, 1972).

WASTE MANAGEMENT SYSTEMS

In the fuel-starved countries of Europe, urban waste incinerators have been integrated into steam plants for electricity generation. In the U.S., however, such wastes have, for the most part, been considered a disposal problem, and their energy has been dissipated through simple incineration or by burial in landfill disposal sites. Environmental problems associated with incinerators or dumping and landfill sites and the diminishing availability of such sites is now resulting in a growing interest in these wastes as an energy resource.

In the U.S. today there are some 3 billion tons of waste accumulated each year. The bulk of this waste is agricultural waste which represents some 2.5 billion tons. Of this, some 2 billion tons is animal manure (Appell, 1971). Of the some 400 million tons of urban waste, 200 million tons is represented by solid wastes collected in garbage trucks. At present we are spending some $3 billion a year to dispose of this material (Grinstead, 1970; Bailie, 1971). In addition, there are the human organic wastes present in municipal sewage systems.

For the most part these wastes are organic material which is primarily cellulose. In other words, it is solar energy stored in organic compounds produced by photosynthesis. On a dry weight basis these wastes would represent some 1 billion tons of cellulose and upon oxidation would yield some 15 quadrillon BTU. This is roughly 25% of our present energy consumption. It also represents 50% of our anticipated oil imports in 1990 (Linden, 1972). Thus, rather than posing an ever increasing disposal problem, these wastes could supply a significant quantity of energy.

With our trend toward urbanization, urban wastes are becoming more concentrated, hence transporation costs for utilizing these wastes are being reduced. A similar trend is occurring with animals. Cattle are being fed in feedlots where from 1,000 to 100,000 cattle are concentrated. It is this high concentration of cattle in feedlots that has exacerbated the animal waste disposal problem. The waste from 100,000 cattle is equivalent to the waste from 200,000 people.

Three approaches have been suggested for converting these wastes into useful energy. One is direct combustion of the material. The second is converting the waste to a fuel by physical processes. The third is conversion to fuel by biological processes.

WASTE AS FUEL

This technology is well developed in Europe, is in use to a small extent in the U.S., and research is being funded by the EPA and the Bureau of Mines (Grinstead, 1970; Bailie, 1971; EPA, 1972). One benefit of this approach is that incineration greatly reduces the bulk of the waste and thus can extend the life of landfill systems severalfold.

In addition to the fuel value of the trash, there is the possibility of recycling the noncombustible materials that comprise some 20 to 30% of the wastes. These materials can be reclaimed from the incinerator or separated beforehand. The market value of this material plus that of the electricity would probably make this approach competitive with present landfill operations that represent the cheapest disposal method (Grinstead, 1970; Bailie, 1971; EPA, 1972). In this respect, note that the economics of solid waste management is determined primarily by the collection system, which involves about 80% of the overall cost. Hence, a doubling of the "disposal" cost would be a small perturbation on the overall costs. Moreover, the environmental savings inherent in recycling operations must be factorized into the economic equation. Utilizing the heat from trash is a form of recycling and also conserves natural resources by displacing an amount of fossil fuels.

However, both from an economic and environmental standpoint, the burning of wastepaper may not be the best approach for this material (Bailie, 1971; Spofford, 1970). Wastepaper represents some 50% of the solid trash, and it can be recycled to produce new paper products. The wastepaper is more valuable as a raw material than as a fuel. To use this approach municipal wastes must be sorted. This technology is well developed and is being funded by private industry, the EPA, the Bureau of Mines, and the Forestry Service (Grinstead 1970, Bailie, 1971; EPA, 1972).

WASTES TO FUEL BY PHYSICAL PROCESS

Two processes have been proposed and tested that convert nonmetallic wastes to usable fuels. One method involves pyrolysis (destructive distillation in the absence of oxygen) (Grinstead, 1970; Spofford, 1970; Malin, 1971) and the other involves the catalytic reduction of organic material with carbon monoxide (Appell, 1971; Grinstead, 1970; Bailie, 1971). Both processes produce low-sulfur fuels with high heating values. In excess of 80% of the heating value of the wastes can be realized in the fuels.

While these processes can be used for urban waste, they also offer great promise in converting the much larger agricultural waste to useful fuel. Another virtue of these processes over direct incineration is that the fuels can be more easily transported to some other use center.

Pyrolysis. The Bureau of Mines still has in operation a research pyrolysis unit built in 1929. Recently, considerable industrial research and development has occurred in this area, and the EPA is also funding such activity (Grinstead, 1970; Malin, 1971). The fuels from this process are a tar-like residue, an oil, and a gas. By a variation of the process, using a catalytic cracking process, up to 70% of energy in solid wastes can be obtained as a gas (Grinstead, 1970).

The present cost for municipal solid waste disposal is one to three dollars per ton in landfill operations and five to ten dollars per ton for incineration. The higher figure for incineration would be expected to be the rule as air pollution control is added to existing facilities. Without recovery of fuel, it is estimated that the pyrolysis process would cost some eight dollars per ton (Malin, 1971). The Bureau of Mines estimates that, after sale of the fuel, the costs would be six dollars per ton for a 500 ton-per-day plant (to supply 200,000 people) and two dollars per ton or less for the 2500 ton-per-day plant (Malin, 1971). Other estimates fall within the same range (Grinstead, 1970). These costs could be reduced by sale of the metals in the solid wastes. Since fuel costs are rising, this approach appears to be economically viable even relative to landfill operations. Again, the costs do not include environmental costs nor the value of the conservation of other fuels.

Current practices for the disposal of animal waste include burial in pits, spreading as fertilizer, and treatment in holding lagoons much like municipal wastes. These approaches have drawbacks in terms of ground and surface water pollution and crop damage from overfertilization. Considerable research is being sponsored in this area by various federal agencies to mitigate these effects (EPA, 1972).

Pyrolysis is also an effective approach for treating animal and agricultural waste. In areas where pollution control is impossible by the above disposal methods, this approach could be an effective solution. As indicated above, the Bureau of Mines estimated a cost of six dollars per ton for a 500 ton-per-day plant. Since 500 tons is equivalent to the manure from 100,000 cattle the cost would be 3 cents per day per steer. Figuring a 6-month feeding cycle in a feedlot for the production of an 800-pound steer, the overall cost would be about 0.7 cent per pound at market time. Moreover, if the manure had to be transported to a central processing facility, this could cost some 30 cents per ton-mile. A 20-mile transport would add six dollars per ton and double the overall cost. With steer prices ranging from 30 to 40 cents per pound, these costs could represent 5% of the marketing price. A dairy cow will produce some 10 liters of milk per day. Transporting the waste 20 miles and processing it would add 0.6 cent per liter of milk. Again this would represent 5 to 10% of the cost of dairy herd management (Mead, 1962). Of course, from these figures one would have to subtract the cost of present waste management practices which range from less than one to a few cents per cow per day (Willrich, 1967; Butchbaker, 1971; Lee and Owens, 1971). Thus it would appear that where it became environmentally necessary, the cost of pyrolysis would not be prohibitive.

One area where the environmental problems are beginning to emerge is the Panhandle area of Texas (Butchbaker, 1971). Feedlots there have a capacity of 2 million head of cattle. Some 3.6 million were fed there in 1970. The cattle density is close to 500 per square mile, and such a region might well find pyrolysis useful.

Catalytic Reduction.   This is another process developed by the Bureau of Mines and is applicable to both urban and agricultural wastes (Appell, 1971). By this process the wastes are mixed with carbon monoxide at a pressure of 4000 pounds per square inch gauge and at temperatures of 350 to 400°C (662 to 752°F) wherein they are converted to a fuel oil. The sulfur content of the oil is only 0.1 to 0.3%, and the oil has a heating value of 15,000 BTU per pound. Carbon monoxide is a reactive catalyst that can be regenerated in the process under proper conditions. Some 80% of the carbon in cellulose material can be converted to oil by this process.

The virtue of this process when compared to pyrolysis is that it produces a single product, oil, whereas pyrolysis produces solid, liquid, and gaseous fuels. At the same time, the requirements for carbon monoxide and high pressure reaction equipment would probably make this a more costly process.

BIOLOGICAL PROCESS

Bacteria are unparalleled chemical processors. Some can utilize simple inorganic molecules and produce exotic organic compounds. Others utilize organic compounds to produce other organic materials, simple inorganic molecules, or both. Groups of different bacteria live in symbiosis with each other.

The first chamber of the cow's stomach contains such a system of bacteria, and the nutrition of the cow is dependent upon these organisms. One of the products of these organisms is methane, and a cow will excrete some 400 liters of methane each day if consuming 10 kilograms of hay per day (Kleiber, 1961). The roughly 100 million cattle in the U.S. thus excrete methane at the rate of some 500 billion cubic feet per year. This is roughly equivalent to our imports of natural gas (about 2% of present consumption) (Linden, 1972). Yet this represents only 10% of the energy in the food consumed by the cows. Another 40% of the energy appears in the manure, and it could be converted to methane in an amount equivalent to some 8% of our consumption.

The above statistics illustrate the sizeable energy content of manure. Moreover, it is important to recall that cattle feed utilizes solar energy with only about 0.1% efficiency and that some agricultural products and algal systems can better this by a factor of 50 or more.

It is the above considerations that have caused schemes relating to the bacterial production of methane to be proposed and studied. One involves converting wastes directly to methane. Another would convert the waste to algae and then to methane. Other systems such as conversion to hydrogen or alcohols are possible. Since the principles are the same, they will not be discussed here.

Methane Production. The production and use of methane from sewage sludge and animal wastes has been a common practice in Europe since 1930 (Oswald and Golueke, 1960; Imhoff, 1946; Gotaas, 1956). The methane is used for all common purposes and, in addition, as fuel for motor vehicles and small motors on farms. The process simply involves the anaerobic decomposition of cellulose materials by bacteria. Both human and animal waste contain the appropriate bacteria for the process. If these wastes are diluted with water and put into a covered tank, a gas is evolved that is about 62% methane, 31% carbon dioxide, 2% hydrogen, and a remaining mixture of nitrogen and other gases (Oswald and Golueke, 1960).

When animal wastes, sewage sludge, and material such as sawdust are employed, some 40% of the available carbon is converted to methane. With cellulose materials such as paper, the yield grows to 63%. Following the digestion of organic wastes, such as manure, some 40% of the original weight of material remains as a humus-like material and a liquor that is rich in nitrogen. In essence, the material left after anaerobic digestion retains the initial fertilizer value of the wastes (Imhoff, 1946).

At the present time, a number of feedlots and municipal sewage plants in this country employ this anaerobic digestion process as a means of treating wastes. However, they do not collect methane, because the additional capital cost would not be offset by the dollar value of the fuel in this country. However, as the price of fuels and energy increase and as the environmental costs are factored into the economic equation, this approach to waste management should reach a competitive position (National Center for Energy Management and Power, 1972). Certainly this approach is worth considering in the Texas Panhandle area discussed above.

Algae-Methane-Algae Systems. A preferred but somewhat more expensive treatment of sewage is aerobic digestion. By this process the sewage is more completely decomposed to carbon dioxide and water by oxygen-requiring bacteria. When these bacteria are coupled with photosynthetic algae, the algae use the carbon dioxide and other nutrients to grow and produce the oxygen required by the bacteria.

Oswald and Golueke have proposed that this process could be used to produce algae (Oswald and Golueke, 1960). The algae would then be decomposed by the anaerobic process to methane that would then be used to generate electrical power. The residue from the anerobic digestion would then be used as nutrient for more algae production. The growth chambers could be enriched with carbon dioxide from the stack of the power plant. In other words, they are proposing an energy farm. They estimate that the cost per kilowatt hour would be three to four times the present costs.

A somewhat similar proposal was submitted by Fredrickson et al. of the University of Minnesota (Fredrickson, 1971). This proposal suggested the possibility of burning the algae directly in a closed-cycle system. It left open the possibility of using an open system, with some of the nutrients being supplied by sewage. They suggested that an overall efficiency of 4% may be obtained, and hence only 160 square miles (0.1% of Minnesota) would supply the electrical needs of the state. They made no cost estimates.

A detailed study sponsored by the NSF-RANN project is underway at the University of Pennsylvania (National Center for Energy Management and Power, 1972). As part of this study the researchers are investigating the possibility of methane production from plants. They suggest that fermentation of seaweeds coupled with the extraction of iodine and potash may be economically competitive. In addition they are considering cultivation of water hyacinth and floating algae. The latter is interesting because floating algae are easy to harvest. The Russians have also expressed interest in such algae (Iycrusalimskiy, 1967).

SUMMARY OF BIOLOGICAL SYSTEMS

A probable limit to the efficiency for conversion of sunlight to chemical energy by plants is some 10%. Even then the conversion is to chemical energy stored in algae or other plants and when electrical power is considered, the overall efficiency will drop to less than 4%. This is to be compared with the 10% or more efficiency possible with physical systems. The economics of the physical systems (some present, some potential) would also appear to be better than the biological algal systems. However, these are only estimates.

At the same time there would appear to be no technological barrier to the development of any of these biological systems and the economic equation, when environmental costs are included, will most likely favor the production of fuel from urban and agricultural wastes by one or more of the processes discussed above. In other words, the value of plants lies in their ability to convert solar energy into food and fiber, and their utility as an energy source lies primarily in the disposal of the resultant cellulose wastes.

The promise of the biological system as a direct energy source will reside in the use of biological reactions and molecules in physical systems. However, it is estimated that such systems lie 20 to 30 years in the future (*New Scientist*, 1972).

CONCLUSIONS

Today, with increasing discussion of the impending energy crisis, more attention is being directed toward the "promise and prospect of solar energy." Much of this interest derives from the consideration that solar energy represents a pollution-free and non-resource-depleting source of energy. It would appear that in the near future, with an adequate program only modestly expensive when compared to today's standards, this source of energy could be utilized in economically competitive fashion. Moreover, it could be utilized without the need for the highly sophisticated technology required for fission or fusion energy application and, hence, be more adaptable to the underdeveloped nations of the world.

However, an energy crisis is not the only problem confronting mankind. Even today, we do not produce sufficient food for the teeming billions who inhabit the earth. Regardless of population control measures, this problem will become more acute in the future. Our primary source of food is the sun.

It is entirely reasonable to suggest that mankind need not diverge from our present day solar economy. The technology is here today, and improvements will require only modestly funded research programs. Solar energy could meet both the food and energy requirements of our present and future population. The state-of-the-art suggests that biological processes should be utilized for their traditional role of food and fiber production and that their contribution to energy or fuels would derive from the management of cellulose wastes. Physical processes should be used to meet energy or fuel needs. The exception to this in the future may be the incorporation of biological molecules in physical systems for direct energy conversion or fuel production.

REFERENCES

Anderson, H. J., and J. H. Anderson, 1966. Mechanical Engineering 88:41.
Anderson, J. H., 1971. The sea plant—a source of power, water, and food without pollution. International Solar Energy Conference at the Goddard

Space Flight Center, Greenbelt, Md., May 12, 1971. *Congressional record,* 117:S16888-S16890.

Appell, H. R., 1971. *Converting organic wastes to oil.* Bureau of Mines Report of Investigation, Report R1-7560.

Arnold, W. F., 1972. *Electronics* 45:67.

Bailie, R. C., 1971. Wasted solids as an energy resource. *Proc. 138th annual meeting of AAAS, symposium on the energy crisis: some implications,* Philadelphia, Dec. 28.

Butchbaker, A. F., 1971. Evaluation of beef feedlot waste management alternatives in livestock waste management and pollution abatement. *Proc. International symposium on livestock wastes, April 19-22, 1971.* American Society of Agricultural Engineers, St. Joseph, Mich., pp. 66–69.

Chemical Week, 1972. Government researchers zero in on applications. 111:57.

Cherry, W. R., 1971. *The generation of pollution-free electric power from solar energy.* Goddard Space Flight Center, Report X-760-71-135.

Eggers, A., 1972. Testimony before Senate Interior Committee, June 7, 1972, as recorded in memo of Senator Gavel, June 12.

Energy Digest, 1972a. NFS to fund cadmium sulfide project in search for "cheap" solar cells. 11:110.

———, 1972b. NFS head views solar energy as most promising alternative 11:1.

Environmental Protection Agency (EPA), 1972. *Environmental protection research catalog, part 1.* EPA Office of Research and Monitoring, Research Information Div., Washington, D.C., Jan., various abstracts, pp. 1–449 to 1–522.

Escher, W. J. D., 1971. Helios-Poseidon: a macro system for the production of storable, transportable energy and foodstuffs from the sun and the sea. American Chemical Society national meeting, Boston, Apr. 9, 1972. *Congressional record,* 117:S17386-S17388.

Ford, N. C., and J. W. Kane, 1971. *Bull. At. Sci.* 27:27.

Fredrickson, A. G., 1971. *Studies on methods to utilize energy income rather than energy capital.* Research proposal submitted to NSF, Univ. of Minn., Minneapolis, June 15.

Gaffron, H., 1960. *Plant Physiol.,* Chap. 4, Part 7, ed. F. C. Steward. New York: Academic Press, 16:114–136.

Glaser, P. E., 1968. *Science* 162:857.

———, 1971. The environmental crisis in power generation and possible future directions. Presented at 39th national meeting of the Operations Research Society of America, Dallas, Tex., May 7, 1971. *Congressional Record,* 117:S10657-S10660.

Gotaas, H. B., 1956. Manure and night soil digesters for methane recovery on farms and in small villages. *Composing,* WHO, Geneva, Switzerland, Chap. 9, pp. 171–193.

Grinstead, R. R., 1970. The new resource. *Environment* 12 (10): 2, Dec.

Heronemus, W. E., 1972. The U.S. energy crisis: some proposed gentle solutions. Joint meeting of ASME and IEE, West Springfield, Mass., Jan. 12, 1972. *Congressional Record,* 118:E1043-1049.

Imhoff, K., 1946. *Sewage Works J.* 18:17.

Industrial Research, 1972. Floating station collects sunlight. 14:32.

Iycrusalimskiy, N. D., 1967. *Controlled biosynthesis.* Joint Publications Research Service, U.S. Dept. of Commerce, JPRS: 41, May 24, p. 143.

Joint Committee on Atomic Energy (JCAE), 1971. Report of JCAE, 92nd Congress, 1st session, Dec., pp. 239–394 (see p. 288).

Jones, L. W., 1971. *Science* 174:367.

———, 1972. A three part series on hydrogen as a fuel. *Chemical and Engineering News* 50(26):14–17, June 26; 50(27):16–19, July 3; 50(28):27–30, July 10.

Kleiber, M., 1961a. *The fire of life.* New York: John Wiley and Sons, pp. 228–341.

———, 1961b. pp. 262–263.

Kok, B., and J. L. P. Van Oorschot, 1954. *Acta Botenica Neerlandica* 3:533.

Lee, H. Y., and T. R. Owens, 1971. Cost of maintaining specific levels of water pollution control for confined cattle feeding operations for the southern high plains. *Proc. International Symposium on Livestock Wastes, Apr. 19–22, 1971.* American Society of Agricultural Engineers, St. Joseph, Mich., pp. 207–208.

Levine, R. P., 1969. *Sci. Am.* 221:58.

Linden, H. R., 1972. The outlook for synthetic fuels. Presented at annual meeting of American Petroleum Institute, Houston, Tex., Mar. 6.

Löf, G., 1955. Cooling with solar energy. In *Proc. on Applied Solar Energy, Phoenix, Ariz.,* Nov. 1–5, pp. 171–190.

Malin, H. M., 1971. *Environ. Sci. Technol.* 5:310.

Mead, S. W., 1962. Dairy cattle management. *Introduction to livestock production,* ed. H. H. Cole. San Francisco: W. H. Freeman and Co., pp. 540–558.

Meinal, A. B., 1971. *A proposal for a joint industry-university-utility task group on thermal conversion of solar energy for electrical power production.* Presentation to Arizona Power Authority, Phoenix, Ariz., Apr. 27, 1971. Reprinted in the *Congressional record,* 117(105):S10655-S10657, July 8.

———, and M. D. Meinel, 1972. *Phys. Today* 25:44.

Meyer, B. S., and D. B. Anderson, 1952. *Plant physiology.* Princeton: Van Nostrand Co., pp. 328–330.

Meyers, J. 1958. *Study of a photosynthetic gas exchanger as a method of providing for the respiratory requirements of the human in a sealed cabin.* School of Aviation Medicine, USAF, Randolph AFB, Tex., Report 58-117.

National Academy of Sciences—National Research Council (NAS-NRC), 1972. Solar cells, outlook for improved efficiency. Ad hoc panel on Solar Cell Efficiency, NAS-NCR, Space Science Board, Washington, D.C.

National Center for Energy Management and Power, 1972. *A status report, National Center for Energy Management and Power.* Univ. of Pensylvania, Philadelphia, May, pp. 16–22.

New Scientist, 1972. Power to the people from leaves of grass. 55:228.

Oak Ridge National Laboratory, 1972a. *An inventory of energy research, vol. 1. Report* ORNL-EIS-72-18V-1.

———, 1972b. pp. 323–336, 555–575.

Oswald, W. J., and C. G. Golueke, 1960. *Biological transformation of solar energy (Advances in applied microbiology, vol. 2.).* New York: Academic Press, pp. 223–262.

Proceedings of the World Symposium on Applied Solar Energy, 1955. Solar house heating—a panel. Held at Phoenix, Ariz., Nov. 1–5. New York: Johnson Reprint Co., pp. 103–158, 1956.

Ralph, E. L., 1971. Large scale solar electric power generation. Presented at Solar Energy Society Conference, Greenbelt, Md., May 10, 1971. *Congressional record,* 117:S10660-S10662.

Rappaport, P., 1959. *R. C. A. Review* 20:373.

Seligen, H., and W. D. McElroy, 1965. *Light: physical and biological action.* New York: Academic Press, pp. 223–236.

Spofford, W. O., 1970. *Environ. Sci. Technol.* 4: 1108.

Tamiya, H., 1957. Mass culture of algae. *Ann. Rev. Plant Physiol.* 8:309–334.

Tybout, R., and G. Löf, 1970. *Natural Resources J.* 10:268.

U.S. Atomic Energy Commission (USAEC), 1968–1970. Civilian nuclear power —a report to the President—1962. Reprinted with appendices in *Nuclear power and related energy problems—1968 through 1970.*

Wassink, E. C., 1953. The efficiency of light-energy conversion in chlorella cultures as compared with higher plants. *Algae culture,* ed. John S. Burlew, Carnegie Institution of Washington, Washington, D.C.

Willrich, T. L., 1967. Disposal of animal wastes. *Agriculture and the quality of our environment,* ed. N. C. Brady, pp. 415–428. Norwood, Mass.: Plimpton Press.

Index